Enhancing the Performance of Ad Hoc Wireless Networks with Smart Antennas

OTHER TELECOMMUNICATIONS BOOKS FROM AUERBACH

Chaos Applications in Telecommunications
Peter Stavroulakis
ISBN: 0-8493-3832-8

Fundamentals of DSL Technology
Philip Golden; Herve Dedieu; Krista S Jacobsen
ISBN: 0-8493-1913-7

IP Multimedia Subsystem: Service Infrastructure to Converge NGN, 3G and the Internet
Rebecca Copeland
ISBN: 0-8493-9250-0

Mobile Middleware
Paolo Bellavista and Antonio Corradi
ISBN: 0-8493-3833-6

MPLS for Metropolitan Area Networks
Nam-Kee Tan
ISBN: 0-8493-2212-X

Network Security Technologies, Second Edition
Kwok T Fung
ISBN: 0-8493-3027-0

Performance Modeling and Analysis of Bluetooth Networks: Polling, Scheduling, and Traffic Control
Jelena Misic and Vojislav B Misic
ISBN: 0-8493-3157-9

Performance Optimization of Digital Communications Systems
Vladimir Mitlin
ISBN: 0-8493-6896-0

A Practical Guide to Content Delivery Networks
Gilbert Held
ISBN: 0-8493-3649-X

Resource, Mobility and Security Management in Wireless Networks and Mobile Communications
Yan Zhang, Honglin Hu, and Masayuki Fujise
ISBN: 0-8493-8036-7

Service-Oriented Architectures in Telecommunications
Vijay K Gurbani; Xian-He Sun
ISBN: 0-8493-9567-4

Testing Integrated QoS of VoIP: Packets to Perceptual Voice Quality
Vlatko Lipovac
ISBN: 0-8493-3521-3

Traffic Management in IP-Based Communications
Trinh Anh Tuan
ISBN: 0-8493-9577-1

Understanding Broadband over Power Line
Gilbert Held
ISBN: 0-8493-9846-0

WiMAX: A Wireless Technology Revolution
G.S.V. Radha Krishna Rao; G. Radhamani
ISBN: 0-8493-7059-0

WiMAX: Taking Wireless to the MAX
Deepak Pareek
ISBN: 0-8493-7186-4

Wireless Mesh Networks
Gilbert Held
ISBN: 0-8493-2960-4

Wireless Security Handbook
Aaron E Earle
ISBN: 0-8493-3378-4

AUERBACH PUBLICATIONS
www.auerbach-publications.com
To Order Call: 1-800-272-7737 • Fax: 1-800-374-3401
E-mail: orders@crcpress.com

Enhancing the Performance of Ad Hoc Wireless Networks with Smart Antennas

Somprakash Bandyopadhyay
Siuli Roy
Tetsuro Ueda

Auerbach Publications
Taylor & Francis Group
Boca Raton New York

Auerbach Publications is an imprint of the
Taylor & Francis Group, an informa business

Published in 2006 by
Auerbach Publications
Taylor & Francis Group
6000 Broken Sound Parkway NW, Suite 300
Boca Raton, FL 33487-2742

© 2006 by Taylor & Francis Group, LLC
Auerbach is an imprint of Taylor & Francis Group

No claim to original U.S. Government works
Printed in the United States of America on acid-free paper
10 9 8 7 6 5 4 3 2 1

International Standard Book Number-10: 0-8493-5081-6 (Hardcover)
International Standard Book Number-13: 978-0-8493-5081-8 (Hardcover)
Library of Congress Card Number 2005057194

This book contains information obtained from authentic and highly regarded sources. Reprinted material is quoted with permission, and sources are indicated. A wide variety of references are listed. Reasonable efforts have been made to publish reliable data and information, but the author and the publisher cannot assume responsibility for the validity of all materials or for the consequences of their use.

No part of this book may be reprinted, reproduced, transmitted, or utilized in any form by any electronic, mechanical, or other means, now known or hereafter invented, including photocopying, microfilming, and recording, or in any information storage or retrieval system, without written permission from the publishers.

For permission to photocopy or use material electronically from this work, please access www.copyright.com (http://www.copyright.com/) or contact the Copyright Clearance Center, Inc. (CCC) 222 Rosewood Drive, Danvers, MA 01923, 978-750-8400. CCC is a not-for-profit organization that provides licenses and registration for a variety of users. For organizations that have been granted a photocopy license by the CCC, a separate system of payment has been arranged.

Trademark Notice: Product or corporate names may be trademarks or registered trademarks, and are used only for identification and explanation without intent to infringe.

Library of Congress Cataloging-in-Publication Data

Bandyopadhyay, Somprakash, 1957-
    Enhancing the performance of ad hoc wireless networks with smart antennas / Somprakash Bandyopadhyay, Siuli Roy, Tetsuro Ueda.
        p. cm.
    Includes bibliographical references and index.
    ISBN 0-8493-5081-6 (alk. paper)
    1. Wireless communication systems. 2. Adaptive antennas. I. Roy, Siuli. II. Ueda, Tetsuro. III. Title.

TK5103.2.B36 2006
621.39'81--dc22   2005057194

Visit the Taylor & Francis Web site at
http://www.taylorandfrancis.com

and the Auerbach Publications Web site at
http://www.auerbach-publications.com

Contents

Preface ix
The Authors xi

1 Introduction 1
   1.1 Ad Hoc Networks: A Preamble 1
   1.2 Characteristics of Ad Hoc Networks 2
   1.3 Some Prospective Usages of Ad Hoc Networks 3
   1.4 Some Research Challenges 5
        1.4.1 Use of Smart Antennas in Ad Hoc Networks 5
        1.4.2 Media Access Control 6
        1.4.3 Routing 6
        1.4.4 Power Conservation 7
        1.4.5 Security 8
   1.5 Performance Evaluation Techniques 8
   1.6 Organization of the Book 9
   References 10

2 The Issues and Challenges in Designing MAC and Routing Protocols 13
   2.1 Media Access Control Techniques 13
        2.1.1 MAC Protocols with Omni-Directional Antennas ... 15
        2.1.2 MAC Protocols with Directional Antennas 19
              2.1.2.1 oRTS/oCTS-Based DMAC 21
              2.1.2.2 DRTS/oCTS-Based DMAC 21
              2.1.2.3 DRTS/DCTS-Based DMAC 21
              2.1.2.4 Other DMAC Protocols 22
        2.1.3 Power-Controlled MAC 24
              2.1.3.1 Power-Control Schemes Using Omni-Directional Antennas 24
              2.1.3.2 Power-Control Schemes Using Directional Antennas 25

2.2 Routing Protocols in Ad Hoc Wireless Networks 27
  2.2.1 Routing Protocols Using Omni-Directional Antennas 27
    2.2.1.1 Reactive Routing Protocols 27
    2.2.1.2 Proactive Routing Protocols 35
  2.2.2 Routing Protocols Using Directional Antennas 38
2.3 Performance Evaluation Techniques 41
  2.3.1 Simulation-Based Evaluation 42
    2.3.1.1 Comparison of Routing Performance 42
    2.3.1.2 Comparison of MAC Performance 44
  2.3.2 Evaluation Using a Testbed 45
References 48

3 Location Tracking and Media Access Control Using Smart Antennas 55
3.1 Introduction 55
3.2 Introduction to Smart Antennas 57
  3.2.1 Features of Smart Antennas 58
  3.2.2 Classification of Smart Antennas 59
  3.2.3 ESPAR: A Smart Antenna for Wireless Ad Hoc Networks 59
  3.2.4 Issues of Smart Antennas in the Context of Wireless Ad Hoc Networks 64
3.3 Location-Tracking Mechanisms for Neighborhood Discovery 65
3.4 Directional Media Access Control Protocols 66
3.5 A Few Assumptions and the Rationales 68
3.6 Performance Evaluation 71
  3.6.1 Simulation Environment 71
  3.6.2 Results 72
3.7 Discussion 76
References 76

4 Location Tracking and Location Estimation of Nodes in Ad Hoc Networks: A Testbed Implementation 79
4.1 Introduction 79
4.2 Location Tracking and Neighborhood Discovery 82
  4.2.1 Formation of the NLST (Neighborhood Link State Table) 82
  4.2.2 Formation of the AST (Angle Signal Table) 83
4.3 Location Estimation 84
  4.3.1 Basic Idea 84
  4.3.2 Location Estimation by a Node Using a Pair of Reference Nodes 86
    4.3.2.1 Synchronization of an Antenna by a Nonprimary Reference Node 87
    4.3.2.2 Formation of the Post-Synchronization Mapping (PM) Table 89
  4.3.3 Estimating Location of a Node Multi-Hop Away from Reference Nodes 90

4.4 Implementation Results 91
    4.4.1 Two-Node Setting: Evaluating Location Tracking 91
    4.4.2 Five-Node Setting: Single-Hop Location Estimations
         with Two Reference Nodes 93
    4.4.3 Five-Node Setting: Multi-Hop Location Estimations
         with Secondary Reference 95
4.5 Error in Location Estimation: A Simulation Study 98
4.6 Discussion 99
References 100

5 A Routing Strategy for Effective Load Balancing Using Smart Antennas 101

5.1 Introduction 101
5.2 System Description 105
    5.2.1 Some Important Definitions 105
    5.2.2 Network Awareness 109
         5.2.2.1 Formation of AST and NANL 110
         5.2.2.2 Formation of ANL 111
         5.2.2.3 Formation of GLST 112
5.3 Network-Aware Routing with Maximally Zone-Disjoint
    Shortest Path 113
    5.3.1 Maximally Zone-Disjoint Shortest-Path Routing 114
    5.3.2 Finding the Maximally Zone-Disjoint Shortest Path:
         An Analysis 116
    5.3.3 Adaptive Route Selection 118
5.4 Maximally Zone-Disjoint Multipath Routing 119
    5.4.1 Multipath Routing Using Omni-Directional and
         Directional Antennas 119
    5.4.2 Selecting Maximally Zone-Disjoint Multipath Routes 123
5.5 Performance Evaluation 124
    5.5.1 Simulation Environment 124
    5.5.2 Impact of Overhead 124
    5.5.3 Evaluation under Static Scenarios 128
    5.5.4 Evaluation under Mobile Scenarios 130
5.6 Discussion 132
References 133

6 Priority-Based QoS Routing Protocols Using Smart Antennas 135

6.1 Introduction 135
6.2 A Few Related Definitions 138
6.3 Priority-Based QoS Routing Using Zone Reservation 139
    6.3.1 Zone Reservation and Route Computation
         by High-Priority Flows 140
    6.3.2 Route Computation and Adaptive Call Blocking
         by Low-Priority Flows 141

　　　　　6.3.2.1　Route Computation without Call Blocking 141
　　　　　6.3.2.2　Route Computation with Call Blocking 142
　　　6.3.3　Performance Evaluation 143
　　　　　6.3.3.1　Effectiveness of Zone-Reservation Protocol 144
　　　　　6.3.3.2　Effectiveness of the Call-Blocking Scheme 145
　6.4　Priority-Based Flow-Rate Control for QoS Provisioning
　　　Using Feedback Control 150
　　　6.4.1　A Control-Theoretic Approach 154
　　　　　6.4.1.1　Some Preliminaries on Proportional-
　　　　　　　　　Integral-Derivative (PID) Control 154
　　　　　6.4.1.2　Priority-Based Flow Control Strategies
　　　　　　　　　Using a PID Controller 155
　　　6.4.2　Priority-Based Flow-Control Scheme Using
　　　　　Directional Antennas 159
　　　　　6.4.2.1　Detecting and Measuring High-Priority
　　　　　　　　　Flow Rates by Other Flows 159
　　　　　6.4.2.2　Feedback Control of Low-Priority Flow Rate 163
　　　6.4.3　Performance Evaluation 164
　　　　　6.4.3.1　Performance of Low-Priority-Flow-Controller
　　　　　　　　　(LPC) 165
　　　　　6.4.3.2　Evaluating the System Performance
　　　　　　　　　in Random Topology 170
　6.5　Service Differentiation in Multi-Hop Intervehicular
　　　Communication 173
　　　6.5.1　Routing in an Unbounded Network 173
　　　　　6.5.1.1　Bounded versus Unbounded Networks 174
　　　　　6.5.1.2　Application-Dependent Route Discovery
　　　　　　　　　Process 175
　　　6.5.2　Implementation of Prioritized Routing Scheme
　　　　　in IVC Scenario 176
　　　　　6.5.2.1　Route Computation and Zone Reservation
　　　　　　　　　by High-Priority Flows 176
　　　　　6.5.2.2　Route Computation and Adaptive Call
　　　　　　　　　Blocking by Low-Priority Flows 177
　6.6　Discussion 178
　References 180

7　Conclusion 183

Index 187

Preface

A mobile ad hoc network (MANET) is a new paradigm of wireless local area networks that enables instantaneous group communications immediately and easily without the aid of any established infrastructure or centralized administration. Usually, the user terminals in ad hoc networks are equipped with omni-directional antennas. However, ad hoc networks with omni-directional antennas normally use a medium access mechanism that wastes a large portion of the network capacity by reserving the wireless media over a large area. To overcome this problem, researchers have proposed to use directional or adaptive antennas that would largely reduce radio interference, thereby improving the utilization of wireless medium and consequently the network throughput. This book will first present an overview of basic Media Access Control (MAC) and routing protocols in ad hoc networks with omni-directional antennas to discuss the issues and challenges. Subsequently, the book will focus on the use of smart antennas in ad hoc networks and discuss the strategies and techniques to be used in designing MAC and routing protocols for improved medium utilization and improved routing performance with effective load balancing. Finally, it will discuss some of the design issues related to priority-based quality-of-service (QoS) routing protocols with smart antennas to illustrate the potential of these antennas vis-à-vis omni-directional antennas in the context of ad hoc networks. Open problems and challenges for ad hoc networks with smart antennas will conclude the book.

The main research work reported in this book has been carried out jointly by ATR Adaptive Communication Research Laboratories in Kyoto, Japan, and the Indian Institute of Management in Calcutta, supported in part by the Telecommunications Advancement Organization of Japan. We would like to express our sincere appreciation to

Dr. Kazuo Hasuike, Dr. Bokuji Komiyama, Dr. Shinsuke Tanaka, and Dr. Sadao Obana for their constant encouragement, guidance, and support. Special thanks go to Mr. Sanjay Chatterjee, Ms. Dola Saha, and Mr. Romit Roy Choudhury for their suggestions and help during the course of our work.

The Authors

Somprakash Bandyopadhyay has a Ph.D. in computer science from Jadavpur University in Calcutta, India, and a B.Tech in electronics and electrical communication engineering from the Indian Institute of Technology in Kharagpur, India. He is currently a professor in the MIS Group of the Indian Institute of Management in Calcutta. He is also the project director of the Ad Hoc Networks Research and Application Group at IIMC. He has more than 22 years of experience in teaching, research, and software development in the Advanced Telecommunications Research Institute, Japan; PricewaterhouseCoopers Limited, Calcutta; Indian Institute of Technology, Kharagpur; Indian Institute of Technology, Bombay; German Research Centre for Artificial Intelligence, Germany; Tata Institute of Fundamental Research, Bombay; and Jadavpur University, Calcutta. He was a fellow of the Japan Trust International Foundation and the Alexander von Humboldt Foundation in Germany.

Siuli Roy has a Ph.D. in computer science from Jadavpur University and an M.C.A. (Master of Computer Applications) from Jadavpur University. She currently works as a principal research engineer at the Indian Institute of Management in Calcutta on the Disaster Management Information Network Project. Prior to that, she worked as a senior research engineer on the ADHOCNET Project at the Indian Institute of Management in Calcutta, sponsored by ATR Adaptive Communications Research Laboratory in Japan. She worked in ATR Adaptive Communications Research Laboratory as a visiting researcher for three months and was involved in the implementation of an ad hoc network testbed using smart antennas.

Tetsuro Ueda received his B.E. degree in electrical and communications engineering and an M.E. degree from Tohoku University and joined NEC Corporation. He has researched channel allocation schemes and worked on the IMT-2000 standardization. He joined KDDI Corporation in 1997 and moved to ATR Adaptive Communications Research Laboratories in 2001. Currently, he is in the YRP Research Center of KDDI R&D Laboratories, Inc. His research interest is wireless ad hoc networks.

Chapter 1

Introduction

1.1 Ad Hoc Networks: A Preamble

Most of the wireless mobile computing applications today require single-hop wireless connectivity to the wired network. This is the traditional cellular network model that supports the current mobile computing needs by installing base stations and access points. In such networks, communications between two mobile hosts rely completely on the wired backbone and the fixed base stations. A mobile host is only one hop away from a base station.

However, at times, no wired backbone infrastructure may be available to support communications among a group of mobile hosts. Also, there might be situations in which setting up fixed access points is not a viable solution due to cost, convenience, and performance considerations. Still, the group of mobile users may need to communicate with each other and share information between them. In such situations, an ad hoc network can be formed.[1,2] An ad hoc network is a temporary network of autonomous nodes that self-organize and self-manage the network without infrastructure support. Nodes in ad hoc networks are computing and communication devices, which can be laptop computers, PDAs, mobile phones, or even sensors. An ad hoc network can be envisioned as a collection of mobile routers, each equipped with a wireless transceiver. The basic assumption in an ad hoc network is that if two nodes willing to communicate are outside the wireless transmission range of each other they may still be able to communicate if other nodes in the network are willing to forward those packets from them. Applications of ad hoc networks include military tactical communication, emergency relief operations, commercial and educational use in remote areas, and in meetings and other situations where the networking is mission oriented or community based.

Usually, the user terminals in ad hoc wireless networks use omni-directional antennas. However, it has been shown that the use of smart directional antennas can largely reduce radio interference, thereby improving the utilization of wireless media and consequently the network throughput.[3-7] Directional antennas have a much higher gain than their omni-directional counterparts, so they significantly reduce the RF power necessary to transmit packets. They can suppress co-channel interference and can therefore enlarge the capacity in terms of node density (more terminals per unit area) in the network. However, in the context of ad hoc networks, it is difficult to find ways to control the direction of such antennas for transmission and the reception in each terminal to achieve an effective multi-hop communication between any source and destination. This difficulty is mainly due to mobility and a lack of centralized control in ad hoc networks. Thus, developing suitable Media Access Control (MAC) and routing protocols in an ad hoc network using directional antennas is a challenging task.

1.2 Characteristics of Ad Hoc Networks

Ad hoc networks have several salient characteristics:[8]

- Infrastructureless, decentralized operation: Usually, ad hoc networks do not rely on any kind of infrastructure support for routing, network management, etc. In other words, ad hoc networks are basically self-organizing and self-managing networks. The lack of centralized control in ad hoc networks requires more sophisticated distributed algorithms to perform all network-related functions.
- Mobility: Mobility causes frequent change in network topology when new nodes join in, some nodes leave or some links break down.
- Multi-hop routing: A node may want to connect to a distant node that is out of its transmission range. Because each node in ad hoc networks can route traffic for the others, multi-hopping

is possible. Multi-hopping is a desirable capability in an ad hoc network because a single-hop ad hoc network does not scale large, thus limiting the communications among the nodes.
- Power-constrained operation: Because nodes can be mobile, they have to rely on battery power, which is a limited resource.
- Heterogeneity: Each node may have different capabilities. In some cases, to be able to connect to an infrastructure-based network (to form a hybrid network), some nodes can communicate with more than one type of network.
- Link asymmetry: In a wireless environment, communication between two nodes may not work equally well in both directions. In other words, even if node n is within the transmission range of node m, the reverse may not be true.
- Bandwidth-constrained, variable capacity links: Wireless links will continue to have significantly lower capacity than their hardwired counterparts. In addition, the realized throughput of wireless communications is often much less than a radio's maximum transmission rate because of the effects of multiple access, fading, noise, and interference conditions, etc. One effect of the relatively low to moderate link capacities is that congestion is typically the norm rather than the exception.

1.3 Some Prospective Usages of Ad Hoc Networks

An ad hoc network complements an infrastructure-based network where infrastructure is not available. Network infrastructure may not be available or viable everywhere, mainly for the following reasons:

- Traffic demand and the forecasted revenue are too low.
- Traffic demand is inconsistent with time (i.e., only existent for a short term).
- It is too costly to build a network infrastructure due to geographical or terrestrial difficulty (e.g., to provide coverage on a highway, in a forest, in remote locations, etc.).

In these cases an ad hoc network can serve as an extension to the network reach. It addresses the need of transporting unplanned or unexpected traffic that is impermanent, or simply, ad hoc based.

The following are some examples where ad hoc networks can be formed without the help of any centralized infrastructural support:

- Community networks are formed in college campuses, city blocks, neighborhoods, and conferences, etc., satisfying intra-community communication needs.[9] A user can be accepted into the network on the spot easily. Community networking allows the formation of local communities (who chat and use a shared whiteboard or who travel in groups) for resource sharing. Community networks tend to be deployed in a very ad hoc manner, wherever resources such as wired Internet access and antenna/radio mounting locations happen to be available. The client node may connect directly to an access point (AP) or use ad hoc routing to leverage other nodes to connect to an AP.
- Enterprise networks are built within corporate environments as low-cost and easy-to-install methods to facilitate the mobility of workers. For example, during a meeting or conference presenters can multicast slides and audio to intended recipients. Wireless multi-hop technology may exist between WLAN APs or between client nodes, or a hybrid ad hoc network can be formed between a WLAN AP network and a client ad hoc network. This is to accommodate the changes caused by human mobility and to allow easy expansion and reconfiguration of network topology (e.g., caused by growth of enterprise staff or indoor renovation).
- Sensor networks[10] can be one of the following: a military sensor network to detect the enemy's movement or chemical/biological weapon detection, an environmental sensing network, a traffic sensor network to monitor traffic congestion in a city, or a surveillance network.
- Emergency response networks can be used for search-and-rescue operations, law enforcement, and disaster relief efforts. As an example of a search-and-rescue operation, each firefighter can carry a communication radio that forms an ad hoc network on site. Also, "breadcrumb routers" can provide connectivity as the firefighters enter the building on fire. As for law enforcement, an ad hoc network can immediately be formed among the patrol cars and handheld radios of police officers at the incident site. Supporting personnel, who later join in, can immediately connect to the ad hoc network that is already formed. In disaster relief operations, ad hoc networks can replace the damaged infrastructure-based network. The ad hoc network addresses the need of immediately deploying a network with high data connectivity on scene.

- Vehicle networks are formed among moving vehicles and land transportation infrastructure (e.g., traffic lights and electronic road signs) and can divert traffic away from congested areas, thus ensuring smooth traffic flow. Intervehicular communication using ad hoc networking is suitable on highways where there is very limited network coverage.[11]

1.4 Some Research Challenges
1.4.1 Use of Smart Antennas in Ad Hoc Networks

There are basically two types of smart antennas used in the context of wireless networks: switched-beam (or fixed-beam antennas) and steerable adaptive array antennas.[12–14] A switched-beam antenna generates multiple predefined fixed directional beam patterns and applies one at a time when receiving a signal. It is the simplest technique and comprises only a basic switching function between separate directive antennas or predefined beams of an array of N antenna elements which are deployed into non-overlapping fixed sectors, each spanning an angle of $360/N$ degrees. Signals will be sensed in all sectors and the antenna is capable of recognizing the sector with the maximum gain. When receiving, exactly one sector, which usually is the one chosen by the sensing process, will collect the signals.

In a steerable adaptive array antenna, which is more advanced than a switched-beam antenna, the beam structure adapts to the radio frequency (RF) signal environment and directs beams toward the signal of interest to maximize the antenna gain, simultaneously depressing the antenna pattern (by setting nulls) in the direction of the interferers.[13] In adaptive array antennas, an algorithm is needed to control the output, i.e., to maximize the signal to interference and noise ratio (SINR). The difference between both kinds of smart antennas is as follows: fixed-beam antennas focus their smartness in the strongest strength signal beam detection, and adaptive array antennas benefit from all the received information within all antenna elements to optimize the output SINR through a weight vector adjustment.

Sundaresan and Sivakumar[15] discussed that superior transmission, reception, and interference suppression capabilities of smart antennas have inspired the consideration of their use in wireless ad hoc networks that are inherently interference limited. However, the capabilities of the antennas can be effectively leveraged only through appropriate changes to higher layer network protocols. Hence, in recent years, several researchers have investigated the following question: What changes need to be made at the MAC layer and above to leverage the unique capabilities of smart antenna technologies?[3,5–7,16–18]

1.4.2 Media Access Control

There are two types of Media Access Control: random or controlled. In random access, all stations compete for a channel in an uncontrolled way (which suits the ad hoc network scenario) and thus collision cannot be avoided. In controlled access, the competition for a channel is controlled (in most cases by a master node) and thus collision can be avoided. Random access suffers from both hidden and exposed terminal problems. The current solution to the hidden terminal problem is to use virtual carrier sensing, which consists of two additional control frames, request to send (RTS) and clear to send (CTS). A terminal that wishes to transmit first sends out an RTS packet. When a destination replies with CTS, all the terminals within its transmission range receive this message and thus the problem of a hidden terminal can be alleviated. The problem with virtual carrier sensing is that many nodes in the vicinity are blocked, and by extending the area in which carrier sensing is effective, although the hidden terminal phenomenon is diminished, the exposed terminal phenomenon increases. One of the solutions to alleviate this problem is to use directional antennas.[4,5,19] This will be discussed in detail in Chapter 2.

1.4.3 Routing

Routing is one of the most important aspects in ad hoc networks because ad hoc network topology frequently changes and multi-hop communication is required. There are two types of routing protocols: proactive and reactive.[1,20–22] Proactive routing protocols are table driven where every node keeps a table of routing information of all the nodes it knows. The routing information is updated periodically. Reactive routing protocols are on demand basis where a route is only formed upon request. Utilizing proactive routing protocols basically gives shorter end-to-end delays because route information is always available and up-to-date as compared to reactive ones. However, the disadvantage is that it consumes more resources (more overhead) in updating the route information. Researchers are now trying to find out the optimal balance between proactive and reactive protocol mechanisms and working on power-aware routing and scalable routing so that routing protocols can still perform under heavy traffic or large numbers of nodes. Provisioning quality of service (QoS) is also being considered as a challenging issue in ad hoc routing. To make an optimal routing decision, QoS routing requires constant updates on link state information such as delay, bandwidth, cost, loss rate, and error rate to make policy decisions, resulting in a large amount of control overhead. This can be prohibitive for bandwidth-constrained ad hoc environments. In addition, the dynamic nature of ad hoc networks makes it extremely difficult to maintain the precise link state information. Finally, even after resource reservation, QoS still cannot be guaranteed due to the frequent disconnections and topology changes. The routing issues and different proposals for routing in ad hoc networks are discussed in detail in Chapter 2.

1.4.4 Power Conservation

Power is a precious resource in mobile devices and networking is one of the most energy-consuming operations. According to an experiment conducted by Kravets and Krishnan,[23] power consumption caused by networking-related activities is approximately 10 percent of the overall power consumption of a laptop computer. This figure rises to 50 percent in handheld devices. The aim of saving power in an infrastructure-based network is to minimize energy consumption in the hosts or nodes. The tactic is to move the communication and computation efforts to the fixed infrastructure, thus keeping the network interface of the devices in sleep state as long as possible. In an ad hoc network every node has to contribute to maintain the network connections. Hence the aim of minimizing the energy consumption of each node is inadequate. An additional aim is to maximize network lifetime. There are two scenarios with regards to energy in ad hoc networks: (1) energy is an expensive, but not a limited resource (batteries can be recharged or replaced easily) and (2) energy is limited, or finite.

The first scenario occurs mostly in community networks and enterprise networks. Thus the objective of the networks belonging to this scenario is to minimize the total energy consumed per packet to forward it from source to destination, without considering the residual energy of individual node. The scenario of finite energy is true in sensor networks. The objective is thus to maximize network lifetime besides conserving energy for individual nodes. Residual energy of individual nodes thus needs to be considered.

1.4.5 Security

Several characteristics of an ad hoc network make it much more difficult to keep its security as compared to the infrastructure-based network:

- Channel vulnerability — broadcast wireless channels allow message eavesdropping and injection easily.
- Node vulnerability — when nodes do not reside in physically protected places, they easily fall under attack.
- Absence of infrastructure — certification/authentication authorities are absent.
- Dynamically changing network topology — this puts security of routing protocols under threat.
- Power and computational limitations — these can prevent the use of complex encryption algorithms.[24]

1.5 Performance Evaluation Techniques

Two approaches can be taken to evaluate the performance of ad hoc networks: by taking measurements from real systems or via modeling and computer simulations.

For the first method, a testbed has to be constructed. The advantage of a testbed is that it is able to reveal problems that cannot be detected by modeling, for example, the communication gray zones problem that has been discovered in AODV implementation.[25] The problem is caused by different transmission ranges for IEEE 802.11b control and data frames, which has been assumed the same in network simulators like NS-2. The disadvantage is that the construction of a testbed is expensive. Moreover, some parameters are difficult to investigate on a real system (e.g., protocol scalability and sensitivity to users' mobility patterns and speeds). Most of the time, measurement results are hardly repeatable. Also a testbed cannot scale to a different size. One of the largest testbeds, the Uppsala University APE testbed, only includes 37 laptop computers in its experiment.[25]

Modeling, on the other hand, makes the study of the network system behavior easy by simply varying the system parameters. Moreover, a large spectrum of network scenarios can be considered. Its disadvantage is associated to its failure to reveal some problems encountered in real implementation. There are two types of modeling: analytical or simulation. Analytical modeling is useful, especially in the study of network or protocol scalability. Simulation studies of scalability are often limited in scope to specific scenarios. There is currently a lack of much-needed theoretical analysis in ad hoc networking partly due to the lack of a common platform to base theoretical comparisons and partly due to the abstruse nature of the ad hoc networking problems. On the other hand, simulation is a standardized, mature, and flexible tool for modeling various protocols and network scenarios. Examples of network simulators are OPNET, NS-2, GloMoSim, and its commercial version, QualNet. The validity of the simulation results based on these tools is now in doubt. Cavin, Sasson, and Schiper[26] have evaluated the flooding algorithm using OPNET, NS-2, and GloMoSim. Their results are different. The differences are not only quantitative (not the same absolute value), but also qualitative (not the same general behavior). This puts some past observation of simulation studies in doubt.

1.6 Organization of the Book

This chapter provides an overview and characteristics of ad hoc networks, their application scenarios, and research challenges. Chapter 2 describes related research efforts in the area of MAC and routing protocols. It begins with a brief review of some well-known existing MAC and routing protocols in ad hoc networks using omni-directional and directional antennas. Chapter 2 also presents a comprehensive survey of existing performance evaluation techniques used by contemporary researchers.

Chapter 3 includes an overview on smart antennas and illustrates the effectiveness of using smart antennas in wireless communication. A novel location-tracking mechanism using directional ESPAR antennas and a receiver-oriented rotational sector-based directional MAC protocol are also presented in this chapter. Our directional routing protocol is tested on this MAC protocol. Simulation results taken on a QualNet network simulator along with the list of parameters are shown here to evaluate the performance of the proposed MAC protocol.

Chapter 4 presents the testbed implementation experience of the proposed location-tracking scheme using directional ESPAR antennas. Besides that, a location estimation method is also proposed, which enables a node to calculate its own position with respect to a pair of fixed reference nodes in a common frame of reference.

A routing strategy for effective load balancing using smart antennas is the subject of Chapter 5. We describe the problem of route coupling in wireless environments and its effect on routing performance. Then the concept of a zone (the key term of our work) in ad hoc networks is illustrated. The rest of the chapter focuses on the operation of network-aware zone-disjoint routing and both single path and multi-path routing schemes are proposed on a unified framework. Simulation results indicating the effectiveness of the proposed routing scheme are also illustrated.

The notion of zone-disjoint routing is further explored in Chapter 6 to propose a priority-based QoS routing protocol for ad hoc networks using directional antennas. This chapter introduces the concept of zone reservation by high-priority flows and eventually temporary call-blocking by low-priority flows, if necessary. A priority-based flow-rate control scheme is also suggested here for QoS provisioning using a feedback control mechanism. The zone reservation and call-blocking mechanism in the context of priority-based QoS provisioning are further extended to use in an unbounded network environment like intervehicular communication (IVC). Simulation results and the set of parameters used in simulation are shown at the end.

Chapter 7 concludes the book with a summary, including the significance of research results. Finally, the chapter offers some recommendations about future work in this research area.

References

1. Johnson, B. and Maltz, D.A., Dynamic Source Routing in Ad Hoc Wireless Networks, in *Proc. Mobile Computing*, T. Imielinski and H. Korth, Eds., Kluwer, 1996.
2. Johnson, D., Routing in Ad Hoc Networks of Mobile Hosts, in *Proc. IEEE Workshop on Mobile Comp. Systems and Appls.*, December 1994.
3. Nasipuri, A. et al., A MAC Protocol for Mobile Ad Hoc Networks Using Directional Antennas, in *Proc. IEEE WCNC*, 2000.
4. Takai, M. et al., Directional Virtual Carrier Sensing for Directional Antennas in Mobile Ad Hoc Networks, in *Proc. ACM MobiHoc*, June 2002.
5. Ramanathan, R., On the Performance of Ad Hoc Networks with Beamforming Antennas, in *Proc. ACM MobiHoc*, October 2001.
6. Roy Choudhury, R. et al., Using Directional Antennas for Medium Access Control in Ad Hoc Networks, in *Proc. ACM MOBICOM*, Atlanta, Georgia, September 2002.
7. Bandyopadhyay, S. et al., An Adaptive MAC Protocol for Wireless Ad Hoc Community Network (WACNet) Using Electronically Steerable Passive Array Radiator Antenna, in *Proc. GLOBECOM*, San Antonio, Texas, November 25–29, 2001.

8. Mukherjee, A., Bandyopadhyay, S., and Saha, D., *Mobile Wireless Networks: Location Management and Routing*, Artech House Books, London, 2003.
9. Mase, K., Sengoku, M., and Shinoda, S., A Perspective on Next Generation Ad Hoc Networks: A Proposal for Open Community Network, *IEICE Trans. Fundamentals*, Vol. E84-A, No. 1, January 2001.
10. Akyildiz, I.F. et al., A survey on sensor networks, *IEEE Communications Magazine*, Vol. 40 No. 8, August 2002, pp. 102–114.
11. Roy, S. et al., Service Differentiation in Multi-hop Inter-Vehicular Communication Using Directional Antenna, in *Proc. VTC spring 2004: IEEE Semiannual Vehicular Technology Conference*, Milan, Italy, May 17–19, 2004.
12. Liberti, J.C. and Rappaport, T.S., *Smart Antennas for Wireless Communications: IS-95 and Third Generation CDMA Applications*, Prentice-Hall, Upper Saddle River, New Jersey, 1999.
13. Lehne, P.H. and Pettersen, M., An Overview of Smart Antenna Technology for Mobile Communications Systems, *IEEE Communications Surveys*, http://www.comsoc.org/pubs/surveys, vol. 2 no. 4, Fourth Quarter 1999.
14. Ueda, T. et al., Evaluating the Performance of Wireless Ad Hoc Network Testbed with Smart Antenna, in *Proc. Fourth IEEE Conference on Mobile and Wireless Communication Networks (MWCN2002)*, September 2002.
15. Sundaresan, K., and Sivakumar, R., A Unified MAC Framework for Ad-hoc Networks with Smart Antennas, in *Proc. ACM International Symposium on Mobile Ad Hoc Networking and Computing (MOBIHOC)*, Tokyo, Japan, May 2004.
16. Nasipuri, A. et al., On-Demand Routing Using Directional Antennas in Mobile Ad Hoc Networks, in *Proc. IEEE International Conference on Computer Communication and Networks (ICCCN2000)*, Las Vegas, October 2000.
17. Roy Choudhury, R. and Vaidya, N., Impact of Directional Antennas on Ad Hoc Routing, in *Proc. 8th Conference on Personal and Wireless Communication (PWC)*, Venice, Italy, September 2003.
18. Bandyopadhyay, S. et al., An Adaptive MAC and Directional Routing Protocol for Ad Hoc Wireless Network Using Directional ESPAR Antenna, in *Proc. ACM Symposium on Mobile Ad Hoc Networking & Computing (MOBIHOC)*, Long Beach, California, October 4–5, 2001.
19. Gyoda, K. and Ohira, T., Beam and Null Steering Capability of ESPAR Antennas, in *Proc. IEEE AP-S International Symposium*, July 2000.
20. Perkins, E. and Bhagwat, P., Highly Dynamic Destination-Sequenced Distance Vector Routing (DSDV) for Mobile Computers, in *Proc. ACM Comp. Comm. Rev. (ACMSICOMM'94)*, Vol. 24, No. 4, October 1994, p. 234.
21. Broch, J. et al., A Performance Comparison of Multi-Hop Wireless Ad Hoc Network Routing Protocols, in *Proc. ACM/IEEE Mobile Comp. and Network*, Dallas, Texas, October 1998.

22. Clausen, T. and Jacquet, P., Optimized Link State Routing Protocol (OLSR), Internet Draft: draft-ietf-manet-olsr-10.txt, May 2003.
23. Kravets, R. and Krishnan, P., Power Management Techniques for Mobile Communication, in *Proc. Fourth Annual ACM/IEEE International Conference on Mobile Computing and Networking (MOBICOM 98)*, October 1998, p. 157.
24. Wang, W., Lu, Y., and Bhargava, B., Intruder Identification in Ad Hoc On-demand Distance Vector Protocol, Department of Computer Sciences technical report, Purdue University, 2002.
25. Lundgren, H. et al., A Large-scale Testbed for Reproducible Ad hoc Protocol Evaluations, in *Proc. WCNC*, 2002.
26. Cavin, D., Sasson, Y., and Schiper, A., On the Accuracy of MANET Simulators, in *Proc. ACM POMC*, Toulouse, France, October 2002.

Chapter 2

The Issues and Challenges in Designing MAC and Routing Protocols

2.1 Media Access Control Techniques

Designing an efficient and effective Media Access Control (MAC) protocol with collision avoidance capability in a mobile ad hoc network is a very challenging task because the network is self-organizing without centralized network control. Moreover, the medium in a wireless network is, by nature, a shared resource where a sender normally uses omni-directional broadcast mode to transmit a message for its intended destination. As a result, other users who are within the transmission range of the sender will also have to "listen" to its message even though they are not the intended receivers of this message. In this context, it is important to ensure a collision-free message communication environment even when multiple senders want to communicate with multiple receivers using the common shared medium. Imagine a situation (Figure 2.1) with seven nodes (I, E, S, D, H, J, and N) where sender E is trying to send a message to receiver I and sender H is trying to send a message to receiver J simultaneously. Let us also

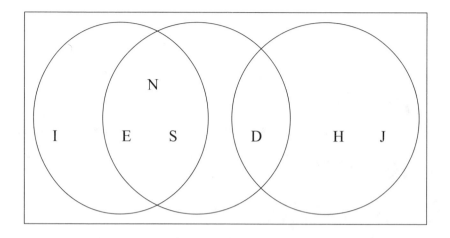

Figure 2.1 Hidden terminal and exposed terminal problems.

assume that I is within the transmission range of E only and J is within the transmission range of H only. In other words, I can only listen to E and not to H. Similarly, J can only listen to H and not to E. This is a conflict-free communication environment where both communications can progress simultaneously without interfering with each other. However, this scenario is rare in the dynamic environment of an ad hoc network. Let us further assume that while E is talking to I, another node S wants to send a message to N. The node N is in the listening range of both E and S. So, there will be collision at receiver N and N will not be able to receive any message from S.

One way to solve the problem is to "sense" the medium before transmitting. So, S will first "sense" the medium to find out the existence of any ongoing communication. Because E is already transmitting data and S is within the transmission range of E, S can "sense" that the medium is busy. So, S will defer its desire to transmit message to N. However, this will give rise to what is known as the exposed terminal problem. Assume that S is transmitting to D. Because E senses an ongoing data transmission (i.e., E is "exposed" to the transmission by S), E remains silent. But E does not know that D is out of its reach. In fact, E could have transmitted to node I because these transmissions would not cause any collision either at D or at I.

A more serious problem is known as the hidden terminal problem. Assume that node S is sending data to node D. A terminal H is "hidden" when it is far away from the data source S but is close to the destination D. Without the ability to detect the ongoing data transmission, H will cause a collision at D if H starts transmitting to J. MAC in this context implements mechanisms to prevent the collision at D by any hidden terminal (such as H).

The main purpose of MAC mechanisms in the context of ad hoc networks is to ensure collision-free communication among nodes under different conditions. However, the collision avoidance schemes using omni-directional antennas waste a large portion of the network capacity by reserving the wireless media over a large area. In other words, when S and D communicate, a large number of nodes in the neighborhood of S and D have to sit idle, waiting for the data communication between S and D to finish. To improve medium utilization, the use of directional antennas is proposed that can largely reduce the radio interference and consequently improve the network throughput. For example, when S and D communicate with directional antennas and S and D are directed toward each other, node E is free to communicate with anyone in its neighborhood except S. Similarly, node H can communicate in all other directions except toward D. However, in the context of ad hoc networks, it is difficult to find ways to control the direction of adaptive antennas for transmission and reception in each terminal to achieve an effective multi-hop communication between any source and destination. This difficulty is mainly due to mobility and lack of centralized control in ad hoc networks. Thus, developing a suitable MAC protocol in an ad hoc network using adaptive antennas is a challenging task.

Another way to improve medium utilization is to use mechanisms of controlling transmission power of senders. For example, a sender may reduce its current transmission range to a minimum value that is sufficient for the successful reception of its message at the intended destination. Medium utilization can be significantly increased because the severity of signal overlapping is reduced. However, adaptive power-control mechanisms introduce additional overheads and problems of route discovery and topology maintenance in ad hoc networks.

2.1.1 MAC Protocols with Omni-Directional Antennas

When nodes in an ad hoc network use omni-directional antennas, it implies that the transmitted signal from any node will spread equally well in all directions within its transmission range. Similarly, as a receiver, a node will receive signals equally well from all directions. Traditionally, an ad hoc network uses an omni-directional antenna for its simplicity and lower cost. However, the use of a directional antenna can largely reduce the radio interference. As a result, utilization of wireless media is improved, which consequently leads to better network throughput. Researchers are also exploring the possibilities of using a low-cost directional antenna in the context of ad hoc networks. In this subsection, MAC protocols with omni-directional antennas will be illustrated. In the next subsection, MAC protocols with directional antennas will be discussed.

Some of the existing MAC protocols can be classified into ALOHA,[1] Carrier Sense Multiple Access (CSMA),[2] Busy Tone Multiple Access (BTMA),[3] Multiple Access with Collision Avoidance (MACA),[4] Media Access Protocol for Wireless LANs (MACAW),[5] Floor Acquisition Multiple Access (FAMA),[6] and Dual Busy Tone Multiple Access (DBTMA).[7] In ALOHA, each user can start a transmission any time without first sensing the channel status. The throughput is affected by frequent collisions. The slotted ALOHA protocol achieves a channel utilization of 36 percent by dividing the channel into time slots. Transmission is allowed to start at the beginning of a time slot only. Slotted ALOHA maintains a higher channel utilization by avoiding collisions in the middle of the data transmission, but it introduces somewhat longer access delay and more complexity due to the need for slot synchronization. CSMA protocols have been used in a number of packet-radio networks in the past.[2] In CSMA, each user senses the carrier before starting the transmission. The throughput is higher than ALOHA because collisions may be avoided. However, the CSMA protocols do not address the hidden terminal problem or the exposed terminal problem, as illustrated in Section 2.1.

The BTMA was the first proposal to combat the hidden terminal problems of CSMA. However, BTMA is designed for station-based networks and requires dividing the channel into a message channel and the busy-tone channel. One of the first protocols for wireless networks based on a handshake between sender and receiver was the SRMA (Split-channel Reservation Multiple Access).[8] It uses the RTS/CTS (request-to-send/clear-to-send) dialogue as a mechanism of handshaking. RTS/CTS dialogue precedes the actual transmission of the data by the stations and allows reserving the channel, so that collisions with other stations are avoided. The MACA and MACAW protocols use a similar scheme in single-channel networks. The protocols based on the RTS/CTS-based handshaking mechanism can reduce the packet loss due to hidden terminals.

In Figure 2.1, before sending the data packets, S sends RTS with proposed duration of data transmission to inform its neighbors about its willingness to start a communication with D. So, all the neighbors of S will become idle (neither transmit nor receive) during this period. If the intended destination D hears the RTS, D replies with CTS to inform its neighbors about its willingness to receive data from S. The neighbors of D that receive *only* CTS cannot transmit (to avoid collision with D's reception), but they can receive from other nodes outside the RTS/CTS boundaries. In some protocols, S issues a data-sending (DS) packet after it receives the CTS packet to assure its neighbors that a successful reservation has been accomplished. This avoids unwanted waiting of nodes receiving only RTS under the condition of unsuccessful negotiation between S and D. After data reception by D is over successfully, it issues an ACK (acknowledgment) to S. The ACK packet, at link level, speeds up packet retransmission that is faster than relying on the slow recovery at the transport layer. Any other nodes overhearing the RTS must be close to node S and therefore should remain silent for a time period long enough so that node S can receive the CTS without any collision. Any other nodes overhearing the CTS must be close to node D and therefore should refrain from transmission for a time period that is long enough for the transmission of the proposed data packet so that node D can receive the data packet without any conflict. A hidden terminal (e.g., node H), which is in the range of node D but out of the range of node S, will hear the CTS but not the RTS. It, therefore, remains silent during the data transmission from node S to node D.

In IEEE's proposed standard for wireless LANs (IEEE 802.11),[9] it describes specifications on the parameters of both the physical (PHY) and MAC layers of the network. The PHY layer, which actually handles the transmission of data between nodes, can use either direct sequence spread spectrum, frequency hopping spread spectrum, or infrared (IR) pulse position modulation. The initial specification of IEEE 802.11 makes provisions for data rates of either 1 Mbps or 2 Mbps, and calls for operation in the 2.4–2.4835 GHz frequency band, which is an unlicensed band for industrial, scientific, and medical (ISM) applications. The current version of IEEE 802.11b operates at 11 Mbps in the 2.4 GHz band and communicates up to 150 feet. IEEE 802.11a operates at 40 Mbps in the 5.8 GHz band. For infrared transmission, it operates at 300–428,000 GHz. Infrared transmissions require absolute line-of-sight links (no transmission is possible outside any simply connected space or around corners), as opposed to radio frequency transmissions, which can penetrate walls.

The basic media access method for 802.11 relies on the distributed coordination function (DCF), which uses carrier sense multiple access/collision avoidance (CSMA/CA) with RTS-CTS-DATA-ACK, as discussed before. A node willing to transmit senses the medium first to determine if the medium is idle. If it is busy, the node waits until transmission stops, and then enters into a random back-off procedure by setting an internal counter. This back-off counter determines the amount of time the node must wait until it is allowed to transmit its packet. During periods when the channel is clear, the node willing to transmit decrements its back-off counter. When the back-off counter reaches zero, the node starts transmitting packets if the medium is still idle. Because the probability that two nodes will choose the same back-off factor is small, this prevents multiple nodes from seizing the medium immediately after completion of the preceding transmission and collisions between packets are minimized.

It is to be noted that carrier sense multiple access/collision detection (CSMA/CD), which is commonly used in wired LANs, is impractical in this context, because WLAN radios are half-duplex and cannot receive while transmitting. Therefore, a collision cannot be detected by a radio while transmission is in progress.

The basic RTS/CTS mechanism described above is supposed to prevent all other nodes in the receiver's range from transmitting. However, CTS packets can still be destroyed by collisions. It has been shown[7] that the probability of CTS packet collision in a multi-hop network that uses the basic RTS/CTS dialogue rules can be as high as 60 percent, when the network operates at high traffic load. To overcome this, the Dual Busy Tone Multiple Access (DBTMA) scheme is proposed. It is based on both the RTS/CTS dialogue and the carrier sensing feature. In particular, carrier sensing is performed by the introduction of two busy tones, which indicate the status of the shared channel in a particular geographical area. This significantly reduces the chances of destruction of the actual data packets due to transmission collisions, which further improves the scheme's utilization. In this protocol, the single common channel is split into two subchannels: a data channel and a control channel. Data packets are transmitted on the data channel, while control packets (RTS, CTS, etc.) are transmitted on the control channel. Additionally, two busy tones are assigned to the control channel: rBT (the receive busy tone, which shows that a node is receiving on the data channel) and tBT (the transmit busy tone, which shows that a node is transmitting on the data channel). The DBTMA scheme significantly improves the performance over the basic RTS/CTS-based schemes. However, the use of a separate channel just to convey the state of the data channel may not be a desirable feature in an ad hoc wireless environment.

FAMA is another proposal to improve basic RTS/CTS-based protocol. FAMA permits a sender to acquire control of the channel in the vicinity of a receiver dynamically before transmitting data packets. The floor acquisition strategy uses an RTS/CTS handshake and is based on a few simple principles: (a) making a sender listen to the channel before transmitting an RTS; (b) implementing a busy-tone mechanism using a single channel and half-duplex radios by making the receiver send a CTS that lasts long enough for the hidden senders to realize that they must back off; and (c) providing priority to those stations who successfully complete a handshake. FAMA is the first single-channel protocol to provide the equivalent functionality of a busy-tone solution and it substantially improves the performance by avoiding collisions.

MACA, MACAW, FAMA, and similar protocols depend on the RTS/CTS dialogue to solve the hidden terminal and exposed terminal problems. However, they do not address the issue of node mobility. The fact that a mobile node did not hear an RTS/CTS dialogue does not indicate that the channel is, indeed, free for use. For example, if a node not hearing the RTS/CTS dialogue moves into the communication range of a receiver, any transmission of the intruder will collide with the ongoing one, and thus affect the channel throughput. This is called an "intruding terminal" problem.[10] To solve this problem, different spread spectrum channels for control messages (i.e., RTS and CTS) and data messages to different nodes have been proposed.[10] Thus, an ongoing data transmission is safeguarded against collisions due to the control or data message from an intruder.

2.1.2 MAC Protocols with Directional Antennas

The RTS/CTS-based scheme using an omni-directional antenna wastes a large portion of the network capacity by reserving the wireless media over a large area. Although a hidden terminal N_R should defer its transmission to protect node R's reception, an exposed terminal N_S unnecessarily defers its transmission despite the fact it would not have interfered with the ongoing S–R communication as shown in Figure 2.2(a). So, RTS/CTS-based omni-directional MAC protocols are very conservative.

However, directional antennas can largely reduce the radio interference, thereby improving the utilization of wireless media.[11-15] It can also eliminate the problem of unnecessary wastage of bandwidth caused by omni-directional RTS (oRTS) and omni-directional CTS (oCTS) using directional RTS (DRTS) and directional CTS (DCTS) as shown in Figures 2.2(b) and 2.2(c).

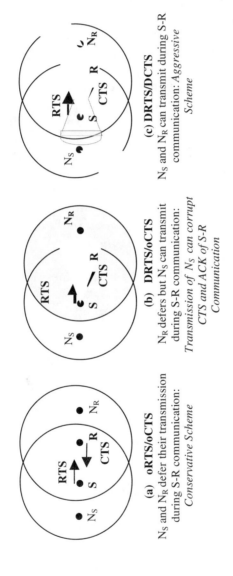

Figure 2.2 Three MAC protocols based on a directional antenna.

2.1.2.1 oRTS/oCTS-Based DMAC

Nasipuri et al. proposed the oRTS/oCTS-based DMAC protocol,[14] where all control packets are transmitted omni-directionally and only data packets are transmitted directionally. Collisions are avoided as in conventional omni-directional MAC algorithms, and the additional benefit is the significant reduction in interference by transmitting and receiving data packets using directional antennas. The key feature of this scheme is a mechanism to determine the direction of the other party's communication. Here, the radio transceiver is assumed to have multiple directional antennas and each node is capable of switching any or all antennas to active or passive modes, known as directional reception capability. An idle node listens to ongoing transmission in every direction. When it receives an oRTS addressed to itself, it can determine the direction of the sender by noting the antenna that received the maximum power of the oRTS packet.[14] Similarly, the sender estimates the direction of the receiver by receiving the oCTS packet. Thus, a receiver is not influenced by other transmissions from other directions. Figure 2.2(a) shows the oRTS/oCTS-based DMAC scheme.

2.1.2.2 DRTS/oCTS-Based DMAC

Ko et al. proposed two DMAC schemes based on DRTS.[13] The first scheme trades off between spatial reuse and collision avoidance by using DRTS and oCTS. Although oCTS helps avoid the collisions from hidden terminals, such as N_R in Figure 2.2(b), DRTS helps improve the spatial channel utilization by eliminating the exposed terminal problem. (N_S is free to attempt its transmission during the S–R communication.) The second scheme uses both DRTS and oRTS to reduce the probability of collisions of control packets in the sender's vicinity caused by the exposed terminal. If there is no ongoing communication in every direction around a sender, then it transmits an oRTS. Otherwise, the sender transmits a DRTS. In both schemes, nodes require external location-tracking support, such as GPS, to determine the direction of the nodes they would like to communicate with. Based on the location of the receiver, the sender may select an appropriate directional antenna to send packets (DRTS and data packets) to the receiver.

2.1.2.3 DRTS/DCTS-Based DMAC

A key question here is how collisions can be avoided with DRTS and DCTS packets. For example, in Figure 2.2(c), S and R are communicating with each other with their directional beams pointing toward each other. Now, if N_R wishes to transmit to R, it simply transmits because N_R did not receive DCTS from node R and is not aware of the S–R communication (deafness problem).[16] But, as R beamforms toward S, it will miss the RTS sent by N_R repeatedly. This goes on until the RTS-retransmit limit has been reached. This amounts to excessive wastage of network capacity in unproductive control packet transmissions. Also, because N_R would increase its back-off interval on each attempt, this phenomenon can result in unfairness as well. Roy Choudhury has also identified another problem with this basic directional virtual carrier sensing scheme known as a hidden terminal due to asymmetry in gain, which occurs when a node, say N_S, waiting in omni-directional idle mode, is not able to sense the ongoing directional communication between S and R as S and R are both out of its omni-directional range.[16] In such a situation, if N_S wants to communicate with R and sends a directional RTS, then it will disturb R's reception from S because the directional gain of N_S is much higher compared to omni-directional gain, thus creating interference to R although R is out of its omni range.

Wang and Garcia-Luna-Aceves observed that the benefit of spatial reuse achieved by a DMAC protocol can outweigh the benefit of a conservative collision avoidance mechanism that sends some omni-directional control packets to silence potential interfering nodes.[17] Their approach uses both DRTS and DCTS and aggressively reuses the channel along the spatial dimension at the cost of increased chance of collisions. In Figure 2.2(c), N_S and N_R can initiate their own transmissions during an S–R communication. It is to be noted that nodes have directional reception capability as discussed previously and thus the transmission from N_R does not cause collisions at node R. Location-tracking support is required for implementing this scheme.

2.1.2.4 Other DMAC Protocols

Multi-hop RTS MAC (MMAC)[16] and Receiver-Oriented Multiple Access (ROMA)[18] are suggested in the context of directional MAC by Roy Choudhury and Bao and Garcia-Luna-Aceves, respectively. The gain of directional antennas is higher than that of omni-directional antennas, and thus they have a greater transmission/reception range.[16] Even if the receiver is within the sender's transmission range, the receiver may not be able to communicate with the sender if its reception range does not include the sender. This is quite possible when the sender transmits directionally, knowing the receiver's location (via GPS), but the receiver tries to receive omni-directionally because it does not know about the transmission attempt from the sender. Therefore, even though data packets can be transmitted over a single hop using a directional antenna at both nodes, it is possible for control packets such as DRTS to take more than one hop. MMAC takes into account this fact and uses multi-hop RTS for delivering DRTS to the receiver over a number of hops.

DMAC protocols proposed by Bao and Garcia-Luna-Aceves are not based on RTS/CTS but use a transmission schedule determined statically based on node identifier and time-slot number.[18] While on-demand medium access schemes determine the communicating pair by exchanging short control signals such as RTS/CTS before each transmission session, scheduled medium access schemes prearrange or negotiate a set of timetables for individual nodes or links. ROMA is such a schedule-based MAC protocol where the communicating nodes are paired with the designated time slots based on the schedule, and thus the transmissions are collision-free.[18]

An adaptive MAC protocol based on an ESPAR (Electronically Steerable Passive Array Radiator) antenna has been proposed.[15] Each node keeps certain neighborhood information dynamically through the maintenance of an angle signal table (AST). Here, each node knows the direction of its neighbors and communication events going on in its neighborhood at that instant of time. Moreover, an appropriate mechanism for null steering of directional antennas in user terminals can help exchanging the neighborhood information in presence of ongoing communication and can improve the medium utilization drastically through overlapping communications in different directions. The angle signal table will also improve the performance of directional routing, because it helps each node to determine the best possible direction of communication with any of its neighbors. In this MAC protocol, initially, when node n wants to communicate with m, it sends an omni-directional RTS to inform all its neighbors, including m, that a communication from n to m has been requested. It also specifies the approximate duration of communication between n and m. All the neighboring nodes of n keep track of this request from node n, whose direction is known to each of them from their respective ASTs. The target node m sends an omni-directional CTS to grant the request and to inform its neighbors that m is receiving data from n. It also specifies the approximate duration of communication. All the neighboring nodes of m keep track of the receiving node m, whose direction is known to each of them from their respective ASTs. After n–m communication begins, other nodes in the neighborhood of both n and m can issue multidirectional RTS and CTS to initiate communications in other directions without disturbing the n–m communication.

Zhuochuan et al. describe the directional version of DBTMA (Dual Busy Tone Multiple Access).[19] It splits a channel into two subchannels. It uses a directional transmitting busy tone, which acts as a directional RTS, and a directional receiving busy tone, which realizes a similar functionality of blocking the corresponding antenna element in the direction from which the omni-directional CTS frame is received.

The protocol presented by Kobayashi and Nakagawa considers that virtual carrier sensing using the RTS and CTS exchange be performed on a sector-by-sector basis rather than on a traditional omni-directional basis.[20] According to their scheme, the network allocation vector (NAV) is maintained separately in each sector (directional NAV), allowing immediate transmission of control packets on those sectors that are clear instead of having to defer the transmission until it is safe to transmit on all sectors at the same time. A problem with this scheme is that if the RTS does not fetch a CTS reply, there is no way for the source node to know whether it is due to the destination not being in one of the free sectors or that it is busy. The source may try to send the RTS in this fashion several times, and if it still does not get a CTS from the destination, the source may assume that the destination is located in one of its sectors in which the RTS was not sent. In that case, the source will wait until all the (off-the-air) OTA waiting times are over and then transmit the RTS omni-directionally. Directional transmission of control packets will provide additional savings of power consumption, though by a small amount, as the control packets are usually much smaller than data and do not consume much power.

2.1.3 Power-Controlled MAC

2.1.3.1 Power-Control Schemes Using Omni-Directional Antennas

Several power-control mechanisms for ad hoc networks have been studied in the context of omni-directional antennas.[21-24] When a node's radio transmission power is controllable, its direct communication range as well as the number of its immediate neighbors is also adjustable. Although higher transmission power increases the transmission range, lower transmission power reduces the collision probability by reducing the number of competing nodes. A wireless signal suffers a variety of degrading effects during propagation, which includes path loss, shadowing, and multipath reflections from surrounding objects.

As a result, the average power of a wireless signal decays exponentially with distance and the envelope of the received signal experiences random fading. Node mobility and multipath reflections also cause the received signal to vary with time. The probability of correctly detecting a received packet depends on the ratio of the signal power of the packet over the total noise and interference power received at the destination from all other transmitters in the network. If the SINR of a received packet falls below a minimum threshold SIR_{min} (which depends on the communication technology), the packet cannot be correctly detected and is considered to have suffered a "collision." So, the reduction of overall interference is a key factor for improving the packet success probability in ad hoc networks.

In the Power Controlled Multiple Access (PCMA) protocol,[23] a source-destination pair uses request power to send (RPTS) and acceptable power to send (APTS) control packets to compute the optimal transmission power based on their received signal strength, which will be used when transmitting data packets. PCMA also uses the busy-tone channel to advertise the noise level the receiver can tolerate. A potential transmitter first senses the busy tone to detect the upper bound of its transmission power for all control and data packets. This transmission power-control approach has been actively studied for other purposes, such as energy saving and topology control. For example, Gomez et al. proposed using the maximum power level for RTS and CTS packets and lower power levels for data packets.[25] This does not increase or decrease the collision probability but nodes can save a substantial amount of energy by using a low power level for data packets. However, this approach has a problem with respect to ACK reception, because EIFS (used to protect ACK) is only effective when data packets are transmitted at full power. The Power Control MAC (PCM) protocol[26] addresses this problem by transmitting data at a reduced power level most of the time but periodically transmits at the maximum power level to inform to its neighboring nodes about the current transmission.

2.1.3.2 Power-Control Schemes Using Directional Antennas

A transmission from a directional antenna reduces interference to all nodes outside of its main lobe. Directional antennas used in reception reduce unwanted interference from surrounding transmitters. Overall, this leads to a higher SINR value for the packet being received, hence increasing the network throughput.

Ramanathan and Nasipuri et al. suggested that, because a directional antenna has a higher gain, a transmitter using a directional antenna requires a lower amount of power to transmit the same distance as would be needed with an omni-directional antenna.[27,28] Transmitting nodes can conserve power by adequately reducing the transmit power when using directional transmissions.

In Ramanathan's work, the performance of a power-control scheme using directional antennas is presented, where the author considered an abstract power-control model that assumes that the transmitter knows the received power at the destination. The main difficulty of implementing power control in an ad hoc network is to enable the source to estimate the required transmit power that will guarantee just the adequate level of SINR of the packet at the receiver. Moreover, in random access channels, there are random variations of the interference level, causing the received SINR for a packet to change randomly even if the transmission power and fading characteristics remain constant.

To maximally utilize the savings in the average power consumption in the network, Nasipuri et al. proposed a power-control scheme that maintains a minimum transmission power level for effective transmission of packets using directional antennas. In that scheme, the RTS and CTS packets are sent with maximum power, but the data packets are transmitted with power control. The RTS/CTS exchange is utilized to determine the power for transmitting the data packet. They assumed that when a destination node receives an RTS packet, it computes the amount by which the SINR of the RTS packet exceeds the SIR_{min} threshold. This information is sent to the source over the CTS packet. The source reduces the power for transmitting the data packet by an amount that is equal to this difference minus a margin δ dB, not exceeding the maximum power level of the transmitter. Hence, the scheme attempts to make the SINR of the received data packet exceed the SIR_{min} threshold by not more than δ dB. The value of δ needs to be chosen carefully, as a small value of δ potentially increases the risk of packet error due to unexpected interference and a large value leads to wastage of power.

Saha et al. studied power control for the purpose of improving SDMA efficiency and to minimize energy consumption.[29] In their paper, a two-level transmit power-control mechanism is proposed to approximately equalize the transmission range R of an antenna operating in omni-directional and directional mode. In other words, if P is the full power used during omni-directional transmission, a reduced power level p will be used during directional transmission so as to equalize the range of transmission approximately in both the cases. This will not only improve the SDMA efficiency but also help to conserve the power of the transmitting node during directional transmission of data. In this scheme, control packets like beacon, RTS, and CTS are omni-directional and use full power P for their transmission. On the other hand, directional transmission of ACK packets and data packets are done with reduced power p.

2.2 Routing Protocols in Ad Hoc Wireless Networks

2.2.1 Routing Protocols Using Omni-Directional Antennas

Routing is a core problem in networks for delivering data from one node to another. Ad hoc networks have special limitations and properties such as *limited bandwidth, highly dynamic topology, link interference, limited range of links*, and *broadcast*. Therefore, routing protocols for wired networks cannot be directly used in wireless ad hoc networks. There are different kinds of routing protocols available for ad hoc wireless networks to date that are either of a reactive or proactive nature. The problem of routing in a network has two components: route discovery and route maintenance. Traditional network routing protocols like Destination Sequenced Distance Vector (DSDV) are proactive, where they maintain a route to all nodes within the network, including those to which no packets are sent. They also react to dynamic topology changes, even if these changes have no effect on the traffic. One major feature of this protocol that it attempts to keep a complete up-to-date routing table for all reachable destinations at all times. In an extremely dynamic environment, where the changes in node connectivity can be very rapid, and, therefore, potentially have a profound effect on the network characteristics, keeping up-to-date links and routes could prove to be detrimental to the routing protocol and the network. Reactive routing protocols, like Dynamic Source Routing (DSR) and Ad Hoc On-demand Distance Vector (AODV), only react when a route is needed between a source and a destination node, and they do not need to try and maintain routes to destinations that they are not communicating with.

2.2.1.1 Reactive Routing Protocols

Dynamic Source Routing (DSR)

Dynamic source routing[30] is based on source routing, where the source specifies the complete path to the destination in the packet header and each node along this path simply forwards the packet to the next hop indicated in the path. It utilizes a route cache where routes it has learned so far are cached. Therefore, a source first checks its route cache to determine the route to the destination. If a route is found, the source uses this route. Otherwise, the source uses a route discovery protocol to discover a route.

In route discovery, the source floods a query packet through the ad hoc network, and the reply is returned by either the destination or another host that can complete the query from its route cache. Each query packet has a unique ID and an initially empty list. On receiving a query packet, if a node has already seen this ID (i.e., duplicate) or it finds its own address already recorded in the list, it discards the copy and stops flooding; otherwise, it appends its own address in the list and broadcasts the query to its neighbors. If a node can complete the query from its route cache, it may send a reply packet to the source without propagating the query packet further. Furthermore, any node participating in route discovery can learn routes from passing data packets and gather this routing information into its route cache.

A route failure can be detected by the link-level protocol (i.e., hop-by-hop acknowledgments) or it may be inferred when no broadcasts have been received for a while from a former neighbor. When a route failure is detected, the node detecting the failure sends an error packet to the source, which then initiates route discovery protocol again to discover a new route. In DSR, no periodic control messages are used for route maintenance.

The major advantage of DSR is that there is little or no routing overhead when a single source or a few sources communicate with infrequently accessed destinations. In such a situation, it does not make sense to maintain routes from all sources to such destinations. Furthermore, because communication is assumed to be infrequent, a lot of topological changes may occur without triggering new route discoveries.

Even though DSR is suitable for the environment where only a few sources communicate with infrequently accessed destinations, it may result in large delay and large communication overhead in highly dynamic environments with frequent communication requirements.[31] Moreover, DSR may have a scalability problem.[32] As the network becomes larger, control packets and message packets also become larger because they need to carry addresses for every node in the path. This may be a problem as ad hoc networks have limited available bandwidth.

Ad Hoc On-Demand Distance Vector (AODV) Routing

The Ad Hoc On-demand Distance Vector (AODV)[33] routing protocol shares DSR's on-demand characteristics in that it also discovers routes on an "as needed" basis via a similar route discovery process. However, AODV adopts a very different mechanism to maintain routing information. It uses traditional routing tables, with one entry per destination. This is a departure from DSR, which can maintain multiple route cache entries for each destination. Without source routing, AODV relies on routing table entries to propagate a route reply back to the source and, subsequently, to route data packets to the destination. AODV uses sequence numbers maintained at each destination to determine freshness of routing information and to prevent routing loops. These sequence numbers are carried by all routing packets.

When a route is needed, a node broadcasts a route request message. The response message is then echoed back once the request message reaches the destination or an intermediate node finds a fresh route to the destination. For each route, a node also maintains a list of those neighbors actively using the route. A link breakage causes immediate link failure notifications to be sent to the affected neighbors. Similar to DSDV, each route table entry is tagged with a destination sequence number to avoid loop formation. Moreover, nodes are not required to maintain routes that are not active. Thus, wireless resources can be effectively utilized. However, because flooding is used for route search, communication overhead for route search is not scalable for large networks. As route maintenance considers only the link breakage and ignores the link creation, the route may become nonoptimal when network topology changes. Subsequent global route search is needed when the route is broken.

An important feature of AODV is maintenance of timer-based states in each node, regarding utilization of individual routing table entries. A routing table entry expires if not used recently. A set of predecessor nodes is maintained for each routing table entry, indicating the set of neighboring nodes that use that entry to route data packets. These nodes are notified with route error packets when the next hop link breaks. Each predecessor node, in turn, forwards the route error to its own set of predecessors, thus effectively erasing all routes using the broken link.

The specification of AODV in Parkins et al.[34] includes an optimization technique to control the route request flood in the route discovery process. It uses an expanding ring search initially to discover routes to an unknown destination. In the expanding ring search, increasingly larger neighborhoods are searched to find the destination. The search is controlled by the TTL field in the IP header of the route request packets. If the route to a previously known destination is needed, the prior hop-wise distance is used to optimize the search.

Associativity-Based Routing (ABR)

Associativity-Based Routing (ABR)[35] also uses a source-initiated method, that is, ABR only maintains routes for sources that actually desire routes. To find a route to the desired destination, ABR also uses query-reply control packets. As in DSR, the source floods a query packet through the ad hoc network to discover a route. However, unlike DSR, the best route is selected by the destination based on the stability of the route and shortest path. A route is considered to be stable if it consists of nodes that have been stationary for a threshold period. The stationary period is measured in terms of "associativity ticks," which are described below.

Each node maintains multiple associativity ticks (one per neighbor), which are initially set to zero. Periodically, each node broadcasts beacons identifying it. These beacons are not newly defined messages but are link-level messages that are normally exchanged between nodes to maintain connectivity. Therefore, no extra overhead is added. Each time a beacon is received, a node increments the associativity tick corresponding to that node from which a beacon is received. If a neighboring node moves out of proximity, the associativity tick that is associated with that neighbor is reset to zero. Therefore, if a mobile host has low associativity ticks with its neighbors, it indicates that the mobile host is in a high state of mobility. On the other hand, if a mobile host has high associativity ticks with its neighbors, it indicates that the mobile host is in the stable (or stationary) state.

In route discovery, the source floods a query packet. As in DSR, a query packet contains an initially empty list, and when receiving a query packet, a node appends its own address in the list. In ABR, however, a query packet also contains the associativity ticks along with other metrics such as hop count. When a node receives a query packet, it appends its address in the list, and it also appends the associativity ticks with its neighbors along with other metrics. The next succeeding node keeps only its upstream neighbor's associativity tick, which is associated with it and erases all others. Therefore, when the query packet reaches the destination, it will contain only the intermediate nodes' addresses and the associativity ticks for that path, together with routing information such as hop count.

After some time period, the destination will know all the possible routes and their qualities. It then selects the best route based on the selection criteria. The most important criterion in selecting a route is the stability of the route indicated by the associativity ticks. Therefore, the destination chooses the route that consists of nodes having high associativity ticks. If more than one route has the same degree of stability, the destination selects the shortest route. The destination then sends a reply packet back to the source. Only the best route selected will be valid, while all other routes will be inactive. When a node moves, unless it is the source, its immediate upstream node erases its route and tries to find an alternate route using the localized query, thereby retaining the control packets locally. Only when necessary, the full query-reply process is performed. Source movement always invokes the full query-reply process.

The major advantage of ABR is that routes selected tend to be long-lived because nodes that have been stationary for some time are less likely to move. This results in fewer route reconstructions, thereby reducing the communication overhead. A possible drawback of ABR is that ABR may have a scalability problem due to the limited available bandwidth, because control packets and message packets need to carry addresses for every node in the path as in DSR.

Signal Stability-Based Adaptive Routing (SSAR)

SSAR[36] also discovers a route only when desired by a source. To find a route to the desired destination, SSAR also uses a query-reply process; the source floods a query packet, and the destination determines the best route and returns a reply packet to the source. The novelty of the SSAR protocol is the use of signal strength as a route selection criterion. The basic idea of SSAR is to select the route, which consists of links with the strong signals because such routes lead to longer-lived routes due to the "buffer zone" effect and consequently require less route maintenance. The buffer zone effect is as follows: if a link between two nodes is a strong link (i.e., a link with strong signal strength), it will have to become a weak link before triggering a route reconstruction. In other words, two nodes that make up the strong link can roam within a certain vicinity of each other without triggering a route reconstruction.

To take the signal strength into account in the routing process, each node periodically broadcasts link-level beacons identifying itself, and upon receiving a beacon, each node records the signal strength at which the beacon was received. If the signal strength is above the threshold, the link is marked as a strong link. Otherwise, it is marked as a weak link. When a node receives a query packet, it propagates the query packet further only if the query packet is received over a strong link and the node has not seen this query packet before. A query packet that is received over a weak link is dropped. When a query packet reaches the destination, it contains the address of each intermediate node. The destination selects the route recorded in the first received query packet because it is probably the shortest path. It then sends a reply packet back to the source along the selected route.

Because only the query packets that are received over the strong links are forwarded further, a query packet may never reach the destination if there is no route that consists of strong links. In this case, the source will not receive any reply packet within some time-out period. When the time-out period expires, the source can wait and try again later, or it may wish to find any route by specially marking the query packet. When a route failure is detected, the node detecting the failure sends an error packet to the source, which then sends a message to erase the invalid route and uses route discovery protocol again to find a new route.

The major advantage of SSAR is that routes selected tend to be long-lived, and this in turn reduces the number of route reconstructions required. Furthermore, SSAR results in low packet loss because strong links are less vulnerable to interference than weak links. A possible drawback of SSAR is that SSAR results in routes with slightly higher hop counts than optimal routing because SSAR prefers routes with strong links that are likely to be between two hosts close to each other.

Stability-Based Routing

In an ad hoc network, the relationship among nodes is based on providing some kind of service, and stability can be defined as the minimal interruption in that service. Irrespective of the routing schemes, frequent interruption in a selected route would degrade the performance in terms of quality of service. Therefore, an important issue is to minimize route maintenance by selecting stable routes, rather than the shortest route. The notion of a path's stability is dynamic and context sensitive. Stability of a path is the span of life of that path at a given instant of time and it has to be evaluated in the context of providing a service. A path between a source and destination is said to be stable if its span of life is sufficient to complete a required volume of data transfer from source to destination. Hence, a given path may be sufficiently stable to transfer a small volume of data between source and destination; but the same path may be unstable in a context where a large volume of data needs to be transferred.

In Stability-based Routing,[37] a notion of link stability and path stability and their evaluation mechanism in the context of dynamic topology changes in an ad hoc network has been proposed. Subsequently, a distributed routing scheme among mobile hosts is proposed in order to find a path between them that is stable in a specific context. They have introduced the notion of link affinity and node affinity in this proposal. The strength of relationship between two nodes over a period of time is defined as node affinity, or affinity. Informally speaking, link affinity $a_{nm}(t)$, associated with a link l_{nm} at time t, is a prediction about the span of life of the link l_{nm} in a particular context. Link affinity $a_{nm}(t)$ at that instant of time is a function of the current distance between n and m, relative mobility of m with respect to n, and the transmission range of n. If transmission ranges of n and m are different, $a_{nm}(t)$ may not be equal to $a_{mn}(t)$. The node affinity or affinity $\eta_{nm}(t)$ between two nodes n and its neighbor m is defined as $\min[a_{nm}(t), a_{mn}(t)]$. The stability of connectivity between n and its neighbor m depends on η_{nm}. The unit of affinity is seconds.

The basic path-searching mechanism given above is the same as in most of the reactive routing with three differences:

- The search is not restricted to find the shortest path; if multiple paths exist between source and destination, the source receives multiple path information from destination in sequence.
- The route reply packet from destination to source would collect the most recent value of affinity a_{mn} for all intermediate nodes m, n,....
- Each path between source and destination is associated with a time delay to estimate the delay associated with that path due to traffic congestion.

When a source initiates a route discovery request, it waits for the route reply until time-out. All the route replies received until time-out are cached at the source. Whenever the source receives the first route reply, it knows the path to the destination and immediately computes its stability η^p_{sd}. If V is the volume of data in the number of packets to be sent to the destination and if B is the bandwidth for transmitting data in packets per second, V/B is the one-hop delay to transmit the data, ignoring all other delay factors. If H_p is the number of hops from the source to the destination in path p, ($H_p \times V/B$) will be the time taken to complete the data transfer. If η^p_{sd} is sufficient to carry this data, the path is selected. Otherwise, the source checks the next path, if available in its cache, for sufficient stability.

When the average mobility of all the nodes in the system is low or volume of data to be communicated between source and destination is low, the chance of route error during data communication with a shortest-path algorithm would be low. A conventional shortest-path routing algorithm would work well in this situation. However, when the average mobility of all the nodes in the system is high or the volume of data to be communicated between source and destination is high, the chance of route error during data communication with a shortest-path algorithm would be high. In this situation, we need to find out a stable path rather than the shortest path for routing. The results in some studies show that the stable-path algorithm reduces route error drastically in all scenarios.[37,38]

Location-Aided Routing

Location-Aided Routing (LAR)[39] uses a global positioning system (GPS) to localize queries to a limited geographic region. This limited region is called the request zone and is expected to hold the current location of the destination. This information is calculated based on the past location of the node and its speed. When a route request is generated at a node S, the request zone of this node is calculated. This information is included in the route request message. Only nodes within the request zone forward the route request. Each node must know its physical location to determine if it is within the request zone or not. If the route discovery, with this initial request zone, time-out and no reply message is received, a new route request is generated using a larger request zone. The maximum request zone is the entire network. It makes sense to dynamically change the request zone after each hop, because we assume that with each hop we will get closer to the destination node. The request zone therefore is calculated and modified in the forwarded request. Advantages of LAR are that it reduces the scope of the route request flood and reduces the overhead of route discovery. Disadvantages of LAR are that the nodes need to know their physical location and their speed to be able to calculate the request zone, and also LAR does not take into account possible existence of obstructions for radio transmissions.

Other than LAR, a number of recent MANET unicast routing protocols use location information. Some of these protocols are the Distance Routing Effect Algorithm for Mobility (DREAM),[40] the Greedy Perimeter Stateless Routing (GPSR) algorithm,[41] the Geographical Routing Algorithm (GRA),[42] the Geographic Distance routing (Gedir) protocol,[43] and the GRID protocol.[44] A review for some of these protocols is provided by Tseng et al.[45] The results presented by Camp et al. show that the use of location information in an ad hoc network significantly improves routing performance of unicast communication.[46]

An adaptive location-aided routing proposed by Nasipuri et al. is effective over a wide range of mobility conditions typical in a MANET. Their combined or hybrid protocol, Adaptive Location-Aided Routing from Mines (ALARM),[47] uses link duration feedback at each node to determine the appropriate forwarding method for data packets. When link durations of nodes on the source route in the packet header are longer than a predetermined threshold, ALARM forwards data packets along this source route (i.e., ALARM uses LAR). When link durations of a node on the source route are shorter than the threshold, that node initiates a directed flood of the data packet toward the destination. This flood is automatically "dampened" when the flooded packets reach nodes that have link durations longer than the threshold. In other words, ALARM may adapt between LAR and directed flooding multiple times as the data packet is routed from the source to the destination. The emphasis is on delivering packets through regions of high network instability. Either the packet reaches the final destination via directed flooding or it picks up the original source route after flooding through a network "hot spot."

2.2.1.2 Proactive Routing Protocols

Destination-Sequenced Distant Vector (DSDV) Routing

DSDV[48] is a distance vector routing protocol that builds and maintains routing tables at each network node. This routing table contains the next hop for, and the total number of hops to, all reachable destinations. As previously stated, DSDV tries to maintain and keep all routing tables completely updated for all connections at all times. It achieves this by periodically sending out updates to all nodes, a network flood type situation. DSDV also uses a hop count and sequence number strategy for routes along its network. This is a combination method that shows how fresh and short a route is. For instance, consider a route A. This route will be considered more favorable than another route, say route B, if either A had a higher sequence number, indicating that it is a fresher route, or if they both had the same sequence number, indicating that route A has a lower hop count. Essentially, the use of sequence numbers helps prevent routing loops when broken links occur. Consider a situation where a route is found to be redundant, and it can no longer be reached. As DSDV broadcasts any change within a network, this redundant information will not remain for very long and potentially waste time and bandwidth. A node along the route would detect the redundant route. Information comprising an infinite hop count would be relayed to the destination node, to enable the destination to increase the sequence number to valid routes by one. Johansson et al. investigated this and the following two main leading routing protocols and came to the conclusion that DSDV is not suitable for wireless ad hoc routing environments, due to a number of failings in the testing carried out.[57] DSDV performs poorly in situations with increased mobility, a metric that helps simulate moving nodes. Because of the way that the DSDV protocol works, it struggles in maintaining valid routes to every node, and lost packets will result. Also, as the network in the simulations grew larger, the protocol had substantial difficulties handling the increased network load as a result of increased updates.

Optimized Link State Routing (OLSR)

The Optimized Link State Routing (OLSR) protocol[49] is another table-driven protocol. There are three main components in this protocol: (a) the neighbor-sensing mechanism, (b) the MPR (multipoint relay) flooding mechanism, and (c) the topology discovery mechanism. In OLSR, nodes exchange messages with other nearby nodes of the network on a regular basis to update topology information on each node. Nodes determine their one-hop neighbors (nodes within one's transmission radius) by transmitting hello messages. Hello messages list neighbors and neighbor quality (whether the link with the neighbor is asymmetric or symmetric and bidirectional). MPRs are selected by a node, among its one-hop neighbors with symmetric, bidirectional links. Each node selects several MPRs from its one-hop neighbors. The MPRs are chosen so that they can broadcast a link state to all of that node's two-hop neighbors. The second technique is link-state reduction. A subset of the total number of link states is flooded throughout the network. OLSR link states only contain the links from a node to its MPRs; the other links are omitted. By these two optimizations, the amount of retransmission is minimized, thereby reducing overhead as compared to link-state routing protocols. Each node will then use this topological information, along with the collected hello messages, to compute optimal routes to all nodes in the network. As such, OLSR is optimized for dense networks with slowly changing topologies. The smaller the subset of MPRs for each node the better the performance. OLSR is designed to work in a fully distributed environment. It sends control messages periodically to its neighboring MPRs, so it can tolerate loss of a few of these messages, because loss of packets is possible due to collision or a hidden terminal problem. OLSR uses hop-by-hop routing. Each hop uses its most recent local information about the next hop.

Fisheye State Routing (FSR)

Fisheye State Routing (FSR)[50] introduces the notion of multilevel fisheye scope to reduce routing update overhead in large networks. Nodes exchange link-state entries with their neighbors with a frequency that depends on distance to the destination. From link-state entries, nodes construct the topology map of the entire network and compute optimal routes. FSR uses the "fisheye" technique proposed by Kleinrock and Stevens, where the technique was used to reduce the size of information required to represent graphical data. The eye of a fish captures with high detail the pixels near the focal point. The detail decreases as the distance from the focal point increases. In routing, the fisheye approach translates to maintaining accurate distance and path quality information about the immediate neighborhood of a node, with progressively less detail as the distance increases. FSR is functionally similar to LS (link state) routing in that it maintains a topology map at each node. The key difference is the way in which routing information is disseminated. In LS, link-state packets are generated and flooded into the network whenever a node detects a topology change.

In FSR, link-state packets are not flooded. Instead, nodes maintain a link-state table based on the up-to-date information received from neighboring nodes and then periodically exchange it with their local neighbors only (no flooding). Through this exchange process, the table entries with larger sequence numbers replace the ones with smaller sequence numbers. The FSR periodic table exchange resembles the vector exchange in DSDV,[48] where the distances are updated according to the time stamp or sequence number assigned by the node originating the update. However, in FSR, link states rather than distance vectors are propagated. Moreover, like in LS, a full topology map is kept at each node and shortest paths are computed using this map.

To reduce the size of update messages without seriously affecting routing accuracy, FSR uses the notion of a multilevel fisheye scope. The scope is defined as the set of nodes that can be reached within a given number of hops. The reduction of routing update overhead is obtained by using different exchange periods for different entries in the routing table. More precisely, entries corresponding to nodes within the smaller scope are propagated to the neighbors with the highest frequency. The rest of the entries are sent out at a lower frequency. As a result, a considerable fraction of link-state entries are suppressed in a typical update, thus reducing the message size. This strategy produces timely updates from near stations, but creates large latencies from far stations. However, the imprecise knowledge of the best path to a distant destination is compensated by the fact that the route becomes progressively more accurate as the packet gets closer to destination. As the network size grows large, a "graded" frequency update plan must be used across multiple scopes to keep the overhead low.

2.2.2 Routing Protocols Using Directional Antennas

Research in the field of ad hoc networks using directional antennas has been confined mostly to issues related to MAC. To determine whether directional antennas are beneficial to ad hoc networks, it is necessary to evaluate the impact of directional antennas on the performance of routing protocols as well.

Although fewer hop routes may be discovered due to the higher transmission range of directional antennas, performing a simple neighborhood broadcast may now require the antenna system to sweep its transmitting beam sequentially over multiple directions. As a result, neighbors of a node receive broadcast packets at different points of time (unlike with omni-directional antennas). Also, sweeping incurs greater delays and can incur higher control overhead, partially negating the advantages derived from reducing the hop count of discovered routes. Work on routing protocols using directional antennas is limited.[47,27,51,52]

Nasipuri et al. have utilized directional antennas for the purpose of on-demand routing.[47] Their primary aim is to minimize routing overhead by intelligently using directional antenna elements for propagating routing information. They have considered a scenario where a source S initially selects a route to destination D through H and I and while the communication is in progress, node I moves away from the transmission range of H on the current path due to mobility and a route error packet is propagated back to the source to initiate a new route discovery process. During this route reconstruction phase, normally an omni-directional antenna causes a query flood that would reach to all the nodes in the network. To limit this flood, the authors proposed two schemes using directional antennas so that RREQ propagation is restricted in the direction from S toward D.

In the first protocol they suggested to transmit the query packets by all the nodes starting from S with the same directional antenna that they have used for the last valid route through D. Each node will propagate the query packet with the same antenna element that they have used for previous valid communication to D. Assuming that D has not moved too far since the time the last route was established, this protocol will restrict the flood of query packets in a region containing the last known location of D. If S does not receive the RREP packet within a suitable time-out period, it decides that the localized search for a route to D has failed and then S will initiate another route discovery, flooding the entire network. A major drawback of this mechanism is that the angular span of the antenna element in S that was used to transmit data packets on the first hop of the last valid route to D may not necessarily include D (that occurs when the first intermediate node on the last valid path to D and the node D do not lie on the same angular span). In that case they have suggested an alternative protocol.

In this scheme they proposed that in addition to the sequence of nodes on a newly discovered route, the header of a route reply packet should also contain the antenna elements to be used at each hop on the path. The source will maintain this information in the route cache. This allows the source to get a rough estimate of the angular location of the destination by simply counting the number of times each antenna element was used and to choose the antenna element with the maximum number of occurrences. Hence, when the route is broken, S may conclude that the possibility of finding a new route to D is highest in the angular span, which has been found to have the maximum number of occurrences. Thus, in the next route discovery phase S will broadcast the query packets to *all* its neighbors with the instruction to rebroadcast it only with the antenna element that has the maximum number of occurrences in the previous path to D. If this search fails to find a path to D, then a networkwide flood will be initiated by S after a suitable time-out period.

Roy Choudhury and Vaidya have suggested a reactive directional routing protocol, which is essentially a modified version of DSR that uses directional antennas (DDSR).[51] When DSR is executed over 802.11 using omni-directional antennas, a single RREQ transmission is received by all the neighbors of the transmitting node. Using the directional MAC protocol (DiMAC) suggested in the paper, broadcasting a packet to all neighbors of a node is achieved through "sweeping." Due to the higher transmission range of directional antennas, the high probability of getting smaller hop paths combined with spatial reuse of directional communication can lead to higher aggregate throughput with DDSR as compared to DSR. Moreover the higher transmission range of directional antennas is often effective in bridging "voids," especially in a sparse network topology.

In their protocol, while performing a sweep, packet transmissions are not preceded by backing off. If the antenna element being used for transmission senses that the channel is idle, it transmits the RREQ. Antenna elements that sense the channel as busy are marked during the first round of transmission. Once the round is completed, DiMAC attempts to transmit the RREQ only over the marked antenna elements (sequentially). If some of the marked antenna elements now sense the channel as idle, the RREQ is transmitted using them. For antenna elements that still sense the channel as busy, DiMAC drops the RREQ. This entire procedure constitutes a single sweep. Roy Choudhury and Vaidya argued that the probability of RREQ collision does not increase much due to the elimination of backing off because the receiver always receives with the antenna that experiences the highest signal strength. Thus, as long as two different signals arrive on different antenna elements, the receiver would always receive only one of them. The authors observed that even if the sweeping delay is high, by using directional transmission a RREQ can be delivered to a faraway node over a single hop.

It has been pointed out that deafness and interference may prevent several nodes from receiving a broadcast RREQ. This may cause the RREQ to reach the destination node through a suboptimal path, increasing route discovery latency as a result. To achieve higher transmission range (narrower beamwidth), the number of antenna elements is required to increase proportionally. This not only increases sweeping delay but worsens the effects of deafness as well. In other words, nodes spend more time in the directional mode while sweeping, increasing the possibilities of being deaf to arriving RREQs. As a result, nodes that are capable of forming the shortest route between the source and destination often do not receive the RREQ. Suboptimal routes are discovered in this scenario. They have suggested an optimization in this context where the RREQ received by the destination may not have traversed the shortest or the least congested path. This optimization requires the destination node to delay sending the RREP by a time duration T calculated from the time it received the first RREQ. This allows the destination node to choose the best among all the routes that arrive within this time T.

Saha et al. present two techniques for routing improvement using directional antennas in MANETs.[52] First, they proposed to use directional antennas to bridge network partitions by adaptively transmitting selected packets over a longer distance, using the capabilities of the directional antenna, yet still transmitting most packets a shorter distance to reduce power consumption and interference. It has been demonstrated through simulations that the DSR protocol when modified using this technique is able to effectively bridge network partitions, and otherwise the protocol is equivalent to the original DSR protocol when no partitions are present. Secondly, they propose a method to use directional antennas to repair routes in use when an intermediate node moves out of wireless transmission range along the route. By using the capability of a directional antenna to selectively transmit packets over a longer distance, they bridge the break in the route caused by the intermediate node's movement, thus reducing packet delivery latency and avoiding dropped packets and additional routing overhead.

2.3 Performance Evaluation Techniques

Protocol performances in ad hoc networks are mostly evaluated on simulators. A number of simulators are currently available, namely NS-2, QualNet, etc., which provide simulation environment for wireless communications. Those simulators tried to incorporate most of the features of wireless communications to characterize real-world wireless communication scenarios. But in real-world scenarios, communication performance in wireless networks can be affected by several factors like terrain, foliage, weather, obstacles, nodal mobility, and the presence electromagnetic radiation, resulting in problems like distance-based attenuation, multipath fading, shadowing, and interference.[53] Simulations do not usually account for all of these effects. In other words, real-world systems face problems that don't occur in simulation. Building real systems using testbeds forces us to handle cases where theoretical models are approximate or do not capture the right details. It is true that unlike simulator experiments, testbed experiments cannot be perfectly reproduced. Interference and radio propagation conditions change between each experiment, and for all practical purposes, they are out of the experimenter's control. However, experimental results are generally repeatable, and running the same experiment many times produces consistent results.[54]

Some simulation-based comparative studies on the performances of currently available MAC and routing protocols will be discussed in the subsequent section. A few initiatives have already been taken by some research groups to evaluate some of the protocols on testbeds. The testbed implementation experiences have also been included in the subsequent section.

2.3.1 Simulation-Based Evaluation

2.3.1.1 Comparison of Routing Performance

The comparative analysis of major unicast routing protocols has been studied extensively through simulation and performance evaluation.[31,55,56,57,58] The performances of DSR and AODV, two prominent on-demand routing protocols for ad hoc networks, have been evaluated by Das et al.[31] A detailed simulation model has been developed to demonstrate the performance characteristics of the two protocols. The general observation from the simulation is that for application-oriented metrics such as delay and throughput, DSR outperforms AODV in less "stressful" situations, i.e., smaller number of nodes and lower load or mobility. AODV, however, outperforms DSR in more stressful situations, with widening performance gaps with increasing stress (e.g., more load, higher mobility). DSR, however, consistently generates less routing load than AODV. The poor delay and throughput performances of DSR are mainly attributed to aggressive use of caching and the lack of any mechanism to determine the freshness of routes when multiple choices are available. The mechanisms to determine freshness of routes will benefit DSR's performance significantly. Because AODV keeps track of actively used routes, multiple actively used destinations also can be searched using a single route discovery flood to control routing load. In general, it was observed that both protocols could benefit from using congestion-related metrics (such as queue lengths) to evaluate routes instead of emphasizing the hop-wise shortest routes and by removing "aged" packets from the network. The aged packets are typically not important for the upper-layer protocol, because they will probably be retransmitted. These stale packets do contribute unnecessarily to the load in the routing layer.

A comprehensive performance evaluation of major proactive and reactive routing protocols has been presented by Das et al.[56] Steady-state performance in terms of fraction of packets delivered, delay, and routing load have been considered as the performance metrics. Even with a packet-level simulation model the essential aspects of the routing protocols are exposed. The key observations are as follows. The proactive, shortest-path protocols provide excellent performance in terms of end-to-end delays and packet delivery fraction at the cost of higher routing load. The on-demand protocols suffer from suboptimal routes as well as worse packet delivery fraction because of more dropped packets. However, they are significantly more efficient in terms of the routing load. The multipath protocol, TORA, did not perform well in spite of maintaining multiple redundant paths. The overhead of finding and maintaining multiple paths and the protocol's sensitivity to the loss of routing packets seem to outweigh the benefits of multiple paths. Also, the end-to-end delay performance is poor because of the loss of distance information. The routing load differentials between all routing protocols reduce with the large number of peer-to-peer conversations in the network. However, the other performance differentials are not affected conclusively.

The results of a detailed packet-level simulation comparing four recent multi-hop wireless ad hoc network routing protocols have been presented.[55] These protocols — DSDV, TORA, DSR, and AODV — cover a range of design choices, including periodic advertisements versus on-demand route discovery, use of feedback from the MAC layer to indicate a failure to forward a packet to the next hop, and hop-by-hop routing versus source routing. They have simulated each protocol in ad hoc networks of 50 mobile nodes moving about and communicating with each other and have presented the results for a range of node mobility rates and movement speeds.

Each of the protocols studied performs well in some cases yet has certain drawbacks in others. DSDV performs quite predictably, delivering virtually all data packets when node mobility rate and movement speed are low, and failing to converge as node mobility increases. TORA, although the worst performer in the experiments conducted in terms of routing packet overhead, still delivered over 90 percent of the packets in scenarios with 10 or 20 sources. At 30 sources, the network was unable to handle all of the traffic generated by the routing protocol and a significant fraction of data packets were dropped. The performance of DSR was very good at all mobility rates and movement speeds, although its use of source routing increases the number of routing overhead bytes required by the protocol. Finally, AODV performs almost as well as DSR at all mobility rates and movement speeds and accomplishes its goal of eliminating source routing overhead, but it still requires the transmission of many routing overhead packets and at high rates of node mobility is actually more expensive than DSR.

2.3.1.2 Comparison of MAC Performance

Huang et al. have done an extensive comparative study[59] of nine existing omni-directional and directional MAC protocols by contrasting their features and evaluating their performances on various network loads and topologies. Among omni-directional MAC protocols they studied IEEE 802.11 MAC DCF, MACA, MACAW, FAMA, and DBTMA. Among directional MAC protocols they compared the performances of MAC/DA1 proposed by Ko et al., MAC/DA2 and MAC/DA2ACK proposed by Nasipuri et al., and DBTMA/DA proposed by Huang et al. The above protocols are grouped into three families rooted from IEEE 802.11 MAC DCF, MACA, and DBTMA. They have compared the characteristic features of different MAC protocols like the antenna type supported by the protocol, whether the protocol supports physical or virtual carrier sensing, the kind of back-off scheme used, whether the protocol is able to solve hidden and exposed terminal problems, whether the protocol splits channels, etc. They have also compared the effectiveness of these MAC protocols in term of spatial reuse. They used an example scenario to evaluate the spatial reuse for each protocol where a node A is transmitting to B and node C is in the neighborhood of B and node E is in the neighborhood of A. C and E both want to start a new communication with D and F, respectively. In this scenario, they tried to figure out how much simultaneous communication each protocol supports. More simultaneous communication indicates better spatial reuse. It has been observed that MAC/DA1 and DBTMA/DA have the highest number of simultaneous communications while 802.11 shows really poor spatial reuse. They validated their observations on the QualNet network simulator and compared the performance in terms of parameters like throughput and end-to-end delay. They have pointed out that application layer metrics are chosen for evaluation of MAC protocols and current ad hoc routing protocols are based on broadcast that normally uses an omni-directional antenna so they chose one-hop traffic or preconfigured static routes for multi-hop flows to avoid the effects of routing protocols to clearly demonstrate the difference between omni and directional MAC protocols. They used CBR traffic to model network load. The simulation results reveal that all directional antenna-based MAC protocols are generally better than 802.11 and also better than MACA. MACA is expected to have better performance when a data transmitter is not blocked by overhearing a CTS frame. Moreover, DBTMA/DA and MAC/DA1 show very similar performances in terms of spatial reuse and throughput but the former usually has longer end-to-end delay. In the case of multi-hop flows the simulation result indicates that most of the directional MAC protocols (DBTMA/DA, MAC/DA1, and MAC/DA2ACK) and MACA have better performance than 802.11 when traffic load is high. MAC/DA2 improves throughput marginally when the ACK frame is used. 802.11 family MAC protocols exhibit significantly longer delay than others, which indicates that using the ACK frame leads to better throughput at the cost of longer delay.

2.3.2 Evaluation Using a Testbed

Over the past few years, some efforts were made to implement an ad hoc network testbed with omni-directional antennas. But only a single testbed project[60] has been reported so far on an ad hoc network using directional antennas in spite of the fact that directional antennas have a number of advantages over omni-directional antennas like spatial reuse, longer transmission and reception range, and better network connectivity.

In the testbed project described by Maltz et al., their primary goal was to build a platform that would enable basic research on the behavior of a real implementation of ad hoc network protocols operating with truly mobile nodes in an outdoor environment.[61,62] Each node in the testbed uses the DSR routing protocol to find and maintain routes to other nodes. They wanted the testbed to operate in an outdoor setting, because many currently envisioned applications of ad hoc networks operate outdoors, and this environment is inherently more unpredictable than an in-building environment. Changes in weather, the motion of cars and pedestrians, and the presence of buildings and hills all affect the propagation of radio signals. Maltz et al. observed significant packet loss due to variation in link quality during multi-hop communication. It has been pointed out that variability in the environment may sometimes create a direct but transient link between source and destination and if those transient links are selected during routing, loss rate increases severely.[61,62] They have argued that better throughput can be achieved if only multi-hop routes can be engaged during the entire course of communication between two far-off nodes.

Chin et al. reported similar behavior while building a MANET testbed.[63] Two routing protocols, namely AODV and DSDV, were implemented on a testbed with five nodes and a four-hop network configuration to study the operational feasibility of existing routing protocols. They reported that the route discovery process of both protocols is fooled by the transient availability of network links to nodes that were more than one hop away. Packets transmitted over a fading channel cause the routing protocol to conclude incorrectly that there is a new one-hop neighbor that could provide a lower metric (hop count) route to even more distant nodes. This can occur even when nodes are stationary; mobility results in even less route stability. They implemented a simple signal strength-based neighbor selection procedure called Powerwave to test their assertion that fading channels and unreliable network links were the cause of the failure of the routing protocols. The result shows that neighbor discovery and the filtering of the reliable (stable) neighbors for routing purposes enables the creation of reliable multi-hop routes.

A similar approach to Powerwave was also taken by Maltz et al.[62] where a program called Macfilter was developed to filter out traffic from unwanted MAC addresses. A novel usage of Macfilter was the emulation of a MANET where multiple nodes could be placed closely together and the signals from neighboring nodes filtered appropriately to give different topologies. The main difference between Macfilter and Powerwave is that Powerwave uses SNR to dynamically determine which IP addresses to filter out whereas Macfilter is statically configured for the topology in question. So, generally, the testbed network included packet filtering software like Macfilter[62] or Powerwave[63] to prevent packets from being transmitted directly from one end node to the other.

Hu and Johnson developed a testbed of eight nodes where each node used DSR routing protocol over 2 Mbps Lucent WaveLAN-II radios.[64] Their objective was to show the performance of an ad hoc network in applications like video conferencing, remote site monitoring, etc. These applications are specially challenging due to rapid changes in routing that may occur with node mobility and due to occurrence of interference and congestion in the shared radio spectrum used by the nodes. Although a number of QoS support schemes have been proposed in this context, those mechanisms introduce complexity into the system and generally increase system overhead. In this project the authors designed and demonstrated a set of three lightweight extensions for DSR in ad hoc networks to support live audio and video streams. Route maintenance in DSR usually attempts a number of retransmissions over a broken link before sending a route error to the source node. Moreover, a new route discovery requires a network round trip between source and destination. Although this total latency is small (typically less than 100 ms) its effect on real-time multimedia streams such as audio and video is undesirable. They have suggested preemptive route maintenance by continuously monitoring the signal-to-noise ratio (SNR) of each received packet by each node. If the SNR value of packets received by a node, say B from A, is found to be less than a certain threshold for a considerable period of time, then the A–B link is likely to break soon. So, before the actual link break occurs, A should send a warning to source node, say S, to find a new route avoiding that link. That will definitely reduce the route maintenance latency.

Lundgren et al. developed an Ad hoc Protocol Evaluation (APE) testbed specifically to experiment with a large number of nodes (37 in this case) to examine scalability and to compare different routing protocols proposed for ad hoc networks.[65] The scenarios are strictly choreographed: after initial "ready-set-go," testbed participants only needed to follow the instructions that appeared on the screen on when and where to move. The nodes moved at a normal walking speed (approx. 1 m/s). A log-driven animation tool called APE-view was written that allows replaying scenarios by positioning the nodes on the screen based on the logged signal quality between node pairs.

De Couto et al. showed that the radio links between the majorities of nodes have substantial loss rates.[54] These loss rates are high enough to decrease forwarding performance but not high enough to prevent existing ad hoc routing protocols from using the links. Link-level retransmission can mask high loss rates, at the cost of substantial decreases in throughput. In many cases, there are longer but higher-quality paths that would afford substantially better end-to-end throughput as well as higher total system capacity. Their testbed is a collection of 18 PCs equipped with 802.11 wireless adapters, distributed around the fifth and sixth floors of the MIT Laboratory for Computer Science so that the resulting network is connected. A simple, proactive routing protocol (a variant of DSDV) runs on the testbed and routes all data traffic. The implementation was done in Click, a modular software router that runs in user level on the testbed nodes.

Toh et al. describe the realization of an ad hoc wireless testbed in an outdoor environment in daytime using four laptops placed in good line-of-sight and each of which included a copy of the TCP/IP/Ethernet protocol suite enhanced with the ABR ad hoc networking software that they had developed.[66] In particular, throughput, end-to-end delay, route discovery time, and the impact of varying source packet size and beaconing intervals are examined. They have suggested the use of localized query/reply packets for route reconstruction/repair in case of unavailability of any intermediate node due to mobility or any other reason during a routing session.

Ramanathan et al. implemented a testbed using a directional antenna and suggested a directional power-controlled MAC, neighbor discovery with beamforming, link characterization for directional antennas, proactive routing, and forwarding on this testbed.[60] They pointed out the necessity of finding the direction of neighbors for proper communication between nodes and suggested three categories of neighbors: N-BF (omni-directional neighbor), T-BF (directional transmit beam and omni-directional receive beam) and TR-BF (directional transmit and receive beams). Two methods were proposed: (1) informed neighbor discovery where a node A will have at least some information about a node B (like geo-location sent by B as updates), which enables A to point its beam toward B, and (2) blind neighbor discovery where a node A has no prior awareness about the existence of a node B. The neighbor discovery mechanism in UDAAN depends on sending and scoring a special periodic control message called heartbeats broadcasted by each node. They have also proposed a routing scheme based on Hazy State Routing Protocol (HSLS). Experimental results of UDAAN show the effectiveness of directional antennas over omni-directional antennas but they have not cited any results showing the accuracy of their neighbor discovery process and performance of routing. Moreover, their directional neighborhood discovery scheme basically relies on a common clock synchronization method using GPS.

References

1. Abramson, N., The ALOHA System — Another Alternative for Computer Communications, in *Proc. Fall Joint Comput. Conf., AFIPS Conf.*, 1970, p. 281.
2. Kleinrock, L. and Tobagi, F.A., Packet Switching in Radio Channels: Part I. Carrier Sense Multiple-access Modes and Their Throughput-delay Characteristics, *IEEE Trans. Communication*, Vol. COM-23, No. 12, December 1975, p. 1400.

3. Tobagi, F.A. and Kleinrock, L., Packet Switching in Radio Channels: Part II. The Hidden Terminal Problem in Carrier Sense Multiple Access Modes and the Busytone Solution, in *IEEE Trans. On Communication*, Vol. COM-23, No. 12, 1975, p. 1417.
4. Karn, P., MACA — A New Channel Access Method for Packet Radio, in *Proc. ARRL/CRRL Amateur Radio 9th Computer Networking Conference, ARRL*, 1990, p. 134.
5. Bharghavan, V. et al., MACAW: A Media Access Protocol for Wireless LANs, in *Proc. ACM SIGCOMM*, 1994, p. 212.
6. Fuller, L. and Garcia-Luna-Aceves, J.J., Floor Acquisition Multiple Access (FAMA) for Packet Radio Networks, in *Proc. ACM SIGCOMM*, 1995, p. 262.
7. Deng, J. and Haas, Z.J., Dual Busy Tone Multiple Access (DBTMA): A Medium Access Control for Multihop Networks, in *Proc. IEEE Wireless Communications and Networking Conference*, New Orleans, Louisiana, September 1999.
8. Tobagi, F.A. and Kleinrock, L., Packet Switching in Radio Channels: Part III. Polling and (Dynamic) Split-channel Reservation Multiple Access Solution, in *IEEE Trans. on Communication*, Vol. COM-24, No. 8, 1976, p. 832.
9. Lough, D.L, Blankenship, T.K., and Krizman, K.J., *A Short Tutorial on Wireless LANs and IEEE 802.11,* The Bradley Department of Electrical and Computer Engineering, Virginia Polytechnic Institute and State University, Blacksburg, April 2000.
10. Ng, M.J., Routing Protocol and Medium Access Protocol for Mobile Ad Hoc Networks, Ph.D. dissertation, Polytechnic University, January 1999.
11. Yum, T.S. and Hung, K.W., Design Algorithms for Multihop Packet Radio Networks with Multiple Directional Antennas Stations, *IEEE Transactions on Communications*, Vol. 40, No. 11, 1992, p. 1716.
12. Zander, J., Slotted ALOHA Multihop Packet Radio Networks with Directional Antennas, *Electronic Letters*, Vol. 26, No. 25, 1990.
13. Ko, Y.B., Shankarkumar, V., and Vaidya, N.H., Medium Access Control Protocols Using Directional Antennas in Ad Hoc Networks, in *Proc. IEEE INFOCOM*, March 2000.
14. Nasipuri, A. et al., A MAC Protocol for Mobile Ad Hoc Networks Using Directional Antennas, in *Proc. IEEE WCNC*, 2000.
15. Bandyopadhyay, S., An Adaptive MAC Protocol for Wireless Ad Hoc Community Network (WACNet) Using Electronically Steerable Passive Array Radiator Antenna, in *Proc. GLOBECOM*, San Antonio, Texas, November 25–29, 2001.
16. Roy Choudhury, R., Using Directional Antennas for Medium Access Control in Ad Hoc Networks, in *Proc. ACM MOBICOM*, Atlanta, Georgia, September 2002.
17. Wang, Y. and Garcia-Luna-Aceves, J.J., Spatial Reuse and Collision Avoidance in Ad Hoc Networks with Directional Antennas, in *Proc. IEEE Globecom*, 2002.

18. Bao, L. and Garcia-Luna-Aceves, J.J., Transmission Scheduling in Ad Hoc Networks with Directional Antennas, in *Proc. 8th Annual International Conference on Mobile Computing and Networking*, Atlanta, Georgia, 2002.
19. Zhuochuan, H., Shen, C., Srisathapornphat, C., and Jaikaeo, C., A Busytone Based Directional MAC Protocol for Ad Hoc Networks, in *Proc. IEEE MILCOM*, Anaheim, California, October 7–10, 2002.
20. Kobayashi, K. and Nakagawa, M., Spatially Divided Channel Scheme Using Sectored Antennas for CSMA/CA–Directional CSMA/CA, in *Proc. PIMRC*, 2000.
21. Agarwal, S., Krishnamurthy, S., Katz, R.H., and Dao, S., Distributed Power Control in Ad Hoc Wireless Networks, in *Proc. PIMRC*, 2001.
22. Kwon, T.J. and Gerla, M., Clustering with Power Control, in *Proc. IEEE MILCOM*, November 1999, p. 1424.
23. Monks, J.P., Varghavan, V., and Hwu, W.W., A Power Controlled Multiple Access Protocol for Wireless Packet Networks, in *Proc. INFOCOM*, 2001.
24. Ramanathan, R. and Rosales-Hain, R., Topology Control of Multihop Wireless Networks Using Transmit Power Adjustment, in *Proc. IEEE INFOCOM*, March 2000, p. 404.
25. Gomez, J., Campbell, A.T., Naghshineh, M., and Bisdikian, C., Conserving Transmission Power in Wireless Ad Hoc Networks, in *Proc. ICNP*, November 2001.
26. Jung, E.S. and Vaidya, N.H., A Power Control MAC Protocol for Ad Hoc Networks, in *Proc. ACM/IEEE MobiCom*, 2002.
27. Ramanathan, R., On the Performance of Ad Hoc Networks with Beamforming Antennas, in *Proc. ACM MobiHoc,* October 2001.
28. Nasipuri, A., Li, K., and Sappidi, U. R., Power Consumption and Throughput in Mobile Ad Hoc Networks Using Directional Antennas, in *Proc. IEEE International Conference on Computer Communication and Networks*, Miami, Florida, October 14–16, 2002.
29. Saha, D. et al., A Power-efficient MAC Protocol with Two-level Transmit Power Control in Ad Hoc Network Using Directional Antenna, in *Proc. 5th International Workshop on Distributed Computing (IWDC)*, December 27–30, 2003.
30. Johnson, B. and Maltz, D.A., Dynamic Source Routing in Ad Hoc Wireless Networks, in *Proc. Mobile Computing*, T. Imielinski and H. Korth, Eds., Kluwer, 1996.
31. Das, S.R., Perkins, C., and Royer, E., Performance Comparison of Two On-demand Routing Protocols for Ad Hoc Networks, in *Proc. IEEE INFOCOM*, Tel Aviv, March 26–30, 2000.
32. Park, V.D. and Corson, M.S., A Highly Adaptive Distributed Routing Algorithm for Mobile Wireless Networks, in *Proc. IEEE INFOCOM*, Kobe, Japan, April 1997.

33. Parkins, C.E. and Royer, E.M., Ad Hoc On Demand Distance Vector Routing, in *Proc. 2nd IEEE Workshop on Mobile Comp. Sys. and Apps.*, February 1999, p. 90.
34. Parkins, C.E., Royer, E.M., and Das, S.R., Ad Hoc On Demand Distance Vector (AODV) Routing, IETF Internet draft, http://www.ietf.org/internet-drafts/draft-ietf-manet-aodv-03.txt, June 1999.
35. Toh, C.K., A Novel Distributed Routing Protocol to Support Ad Hoc Mobile Computing, in *Proc. IEEE International Phoenix Conference on Computer and Communications*, 1996.
36. Dube, R. et al., Signal Stability-based Adaptive Routing for Ad Hoc Mobile Networks, Institute of Advanced Computer Studies, Department of Computer Science, Technical Report CS-TR-3646, UMIACS-TR-96-34, University of Maryland, College Park, August 1996.
37. Paul, K. et al., A Stability-based Distributed Routing Mechanism to Support Unicast and Multicast Routing, in *Ad Hoc Wireless Network, Computer Communications* (Elsevier Science), Vol. 24, December 2001, p. 1828.
38. Paul, K., Roy Choudhury, R., and Bandyopadhyay, S., Survivability Analysis Ad Hoc Wireless Network Architecture, in *Proc. Second International Workshop on Mobile and Wireless Communication Networks (LNCS 1818, Springer Verlag)*, Paris, France, May 2000.
39. Ko, Y.B. and Vaidya, N.H., Location-aided Routing in Mobile Ad Hoc Networks, in *Proc. ACM/IEEE MobiCom*, October 1998.
40. Basagni, S. et al., A Distance Routing Effect Algorithm for Mobility (DREAM), in *Proc. Fourth Annual ACM International Conference on Mobile Computing and Networking (MobiCom 1998)*, 1998, p. 76.
41. Karp, B. and Kung, H. T., GPSR: Greedy Perimeter Stateless Routing for Wireless Networks, in *Proc. Sixth Annual ACM International Conference on Mobile Computing and Networking (MobiCom 2000)*, 2000, p. 243.
42. Jain, R., Puri A., and Sengupta, R., Geographical Routing Using Partial Information for Wireless Ad Hoc Networks, in *Proc. IEEE Personal Communications*, February 2001, p. 48.
43. Stojmenovic, I. and Lin, X., Loop-free Hybrid Single-path/Flooding Routing Algorithms with Guaranteed Delivery for Wireless Networks, in *IEEE Transactions on Parallel and Distributed Systems*, Vol. 12, No. 10, 2001, p. 1023.
44. Liao, W.H., Tseng, Y.C., and Sheu, J.P., Grid: A Fully Location-aware Routing Protocol for Mobile Ad Hoc Networks, in *Telecommunication Systems*, Vol. 18, No. 1, 2001, p. 37.
45. Tseng, Y.C. et al., Location Awareness in Ad Hoc Wireless Mobile Networks, in *Computer*, Vol. 34, No. 6, 2001, p. 46.
46. Camp, T. et al., Performance Comparison of Two Location-based Routing Protocols for Ad Hoc Networks, in *Proc. 21st Annual Joint Conference of the IEEE Computer and Communications Societies (Infocom 2002)*, 2002, p. 1678.

47. Nasipuri, A. et al., On-demand Routing Using Directional Antennas in Mobile Ad Hoc Networks, in *Proc. IEEE International Conference on Computer Communication and Networks*, Las Vegas, Nevada, October 2000.
48. Perkins, E. and Bhagwat, P., Highly Dynamic Destination-Sequenced Distance Vector Routing (DSDV) for Mobile Computers, in *Proc. ACM Comp. Comm. Rev.*, Vol. 24, No. 4 (ACMSICOMM '94), October 1994, p. 234.
49. Clausen, T. and Jacquet, P., Optimized Link State Routing Protocol (OLSR), Internet Draft: draft-ietf-manet-olsr-10.txt, May 2003.
50. Pei, G., Gerla, M., and Chen, T.W., Fisheye State Routing: A Routing Scheme for Ad Hoc Wireless Networks, in *Proc. IEEE International Conference on Communication*, New Orleans, Louisiana, June 2000.
51. Roy Choudhury, R. and Vaidya, N., Impact of Directional Antennas on Ad Hoc Routing, in *Proc. 8th Conference on Personal and Wireless Communication (PWC)*, Venice, September 2003.
52. Saha, A. and Johnson, D.B., Routing Improvements Using Directional Antennas in Mobile Ad Hoc Networks, in *Proc. Globecom 2004*, Dallas, Texas, November 2004.
53. Streenstrup, M.E., Neighbor Discovery among Mobile Nodes Equipped with Smart Antennas, www.wireless.kth.se/adhoc03/Proceedings/mssession6-1.pdf.
54. De Couto, D.S.J., Aguayo, D., Benjamin, A.C., and Robert, M., Effects of Loss Rate on Ad Hoc Wireless Routing, MIT Laboratory for Computer Science technical report, MIT-LCS-TR-836, March 2002.
55. Broch, J et al., A Performance Comparison of Multi-hop Wireless Ad Hoc Network Routing Protocols, in *Proc. ACM/IEEE Mobile Comput. and Network*, Dallas, Texas, October 1998.
56. Das, S.R., Castañeda, R., Yan, J., and Sengupta, R., Comparative Performance Evaluation of Routing Protocols for Mobile Ad Hoc Networks, in *Proc. of IEEE IC3N'98*, Lafayette, Louisiana, October 1998, p. 153.
57. Johansson, P. et al., Scenario-based Performance Analysis of Routing Protocols for Mobile Ad Hoc Networks, in *Proc. of ACM/IEEE MOBICOM'99*, Seattle, Washington, August 1999, p. 195.
58. Lee, S.J., Gerla, M. and Toh, C.K., A Simulation Study of Table-driven and On-demand Routing Protocols for Mobile Ad Hoc Networks, in *IEEE Network*, Vol. 13, No. 4, July 1999, p. 48.
59. Huang, Z. and Shen, C.C., A Comparison Study of Omnidirectional and Directional MAC Protocols for Ad Hoc Networks, in *Proc. Globecom*, 2002.
60. Ramanathan, R. et al., Ad Hoc Networking with Directional Antennas: A Complete System Solution, in *Journal of Selected Areas in Communications*, Vol. 23, No. 3, March 2005, p. 496.

61. Maltz, D.A., Broch, J., and Johnson, D. B., Experiences Designing and Building a Multi-hop Wireless Ad Hoc Network Testbed, CMU-CS-99-116, Carnegie Mellon University, School of Computer Science, March 1999.
62. Maltz, D.A., Broch, J., and Johnson, D.B., Quantitative Lessons from a Full-scale Multi-hop Wireless Ad Hoc Network Testbed, in *Proc. IEEE Wireless Communications and Networking Conference*, September 2000.
63. Chin, K.W. et al., Implementation Experience with MANET Routing Protocols, in *Proc. ACM SIGCOMM Computer Communication Review*, Vol. 32, No. 5, November 2002, p. 49.
64. Hu, Y.C. and Johnson, D.B., Design and Demonstration of Live Audio and Video over Multihop Wireless Ad Hoc Networks, in *Proc. MILCOM 2002 IEEE Military Communications Conference*, IEEE, Anaheim, California, October 2002.
65. Lundgren, H. et al., A Large-scale Testbed for Reproducible Ad hoc Protocol Evaluations, in *Proc. WCNC*, 2002.
66. Toh, C.K. et al., Experimenting with an Ad Hoc Wireless Network on Campus: Insights and Experiences, in *ETRICS Performance Evaluation Review*, Vol. 28, No. 3, December 2000, p. 21.

Chapter 3

Location Tracking and Media Access Control Using Smart Antennas

3.1 Introduction

Ad hoc networks with omni-directional antennas use an RTS/CTS-based floor reservation scheme that wastes a large portion of the network capacity by reserving the wireless media over a large area. Consequently, lots of nodes in the neighborhood of transmitter and receiver have to sit idle, waiting for the data communication between transmitter and receiver to finish. To alleviate this problem, researchers have proposed to use directional (fixed or adaptive) antennas that direct the transmitting and receiving beams toward the receiver and transmitter node only. This would largely reduce radio interference, thereby improving the utilization of wireless media and consequently the network throughput.[1-11] As shown in Figure 3.1, while node n communicates with node m using an omni-directional antenna, nodes p and r have to sit idle. However, with directional beamforming, while node n communicates with node m, nodes p and r can communicate with nodes q and s, respectively, improving the medium utilization or the SDMA (Space Division Multiple Access) efficiency drastically.

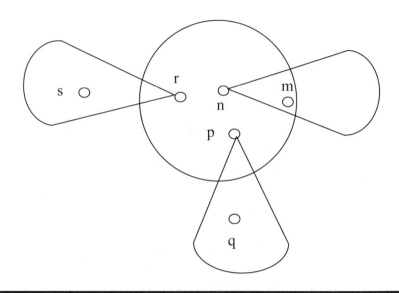

Figure 3.1 Improving SDMA efficiency with directional antennas.

To fully exploit the capability of a directional antenna, it is necessary for each node to know the information of the neighboring nodes (such as node-ID, direction, link quality, etc.) beforehand. Node p can initiate communication with q only if the direction from p to q is not in the same direction as p to m or p to n. Thus, whenever a source and a destination node are engaged in a communication, all the neighbors of the source and destination nodes should know the direction of communication so that they can initiate new communication in other directions, thus preventing interference with ongoing data communication between the source and destination. In other words, a node should know how to set its transmission direction to transmit a packet to its neighbors to implement effective MAC and routing protocols in this context. So, it becomes imperative to have a mechanism at each node to track the locations of its neighbors. However, this location-tracking mechanism in the context of wireless ad hoc networks with directional antennas is a serious problem, because it incurs a lot of control overhead. In this chapter, we have included an overview of smart antennas. The concepts and characteristic features of smart antennas and their classifications and working principles are discussed in detail. This chapter also includes an example of a smart directional antenna called an ESPAR antenna developed by ATR-ACR (Adaptive Communications Research Laboratories) in Japan. ATR-ACR has developed a wireless ad hoc network testbed using the ESPAR antenna to experimentally assess the performances of MAC and routing protocols when directional and adaptive beam patterns are used for data communication. In the next section, we will illustrate a receiver-centric approach for location tracking and MAC protocol. To track the location of its neighbor, each node n periodically collects its neighborhood information and forms an angle signal table (AST). Based on the AST, a node n knows the direction of node m and controls the medium access during transmission-reception. The performance evaluation on a QualNet network simulator[12] indicates that the said protocol is highly efficient with an increasing number of communications and with an increase in data rate. The one-hop MAC throughput of the protocol is 1.8 times as compared to that of IEEE 802.11.

3.2 Introduction to Smart Antennas

To increase the capacity, the same frequencies can be reused between cells, where the distance between cells is far enough not to interfere with each other. Microcell is the popular technique to improve the capability in cellular systems in this context. Because wireless ad hoc networks assume the smaller reuse distance between each coverage area of mobile stations, a wireless ad hoc network is an efficient system in the point of frequency utilization. Although wireless ad hoc networks use the packet forwarding technique for the other mobile users, many routes occupy the same communication area. Consequently, even if the frequency reuse technique increases the total capacity of ad hoc networks, it is required to study some technique to avoid the co-channel interference from the other users and for more routes to coexist within the same area.

Moreover, the radio propagation condition between each mobile subscriber is very complex. There is significant variation in the received signals with the mobility of the hosts. The received signal of each node consists of line-of-sight paths and multiple reflected paths, which are combined as fading models with Rayleigh distribution or Ricean distribution. Even for stationary nodes, any change in the line-of-sight component or any reflected component will change the channel quality. A physical-layer multirate capability, such as IEEE 802.11a, IEEE 802.11b, and IEEE 802.11g, etc., will change the data rate that is reasonable to the channel conditions and SINR (signal to interference and noise radio) value. However, within the same frequency area, even if the data rate is low and consequently the required SINR is small, it is very difficult to keep that SINR level, due to co-channel interference. Moreover, the above-mentioned direct components and reflected radio paths that are caused by some obstacles, such as buildings and houses, degrade the signal quality. Concretely, delay spread, which is the difference between arrival times of direct and reflected paths, creates the channel distortion as ISI (inter symbol interference). Under the multiuser scenario, not only the reflected paths from the user itself, but also the direct and reflected paths from the other users should be taken into account. The smart antenna is one of the most promising techniques for increasing the capacity and improving link quality in severe multipath propagation.

3.2.1 Features of Smart Antennas

The most interesting feature of smart antennas is that they can increase the capacity through the beam and null controls. Usually, mobile terminals communicate using an omni-directional antenna. This antenna radiates the signal in all directions. The signal would create interference to the other mobile nodes except the target mobile terminal. Therefore, only one source and destination pair could use the same frequency coverage area. However, a smart antenna steers its beam only in the direction of the target destination node. In addition, it creates nulls to the directions except the target mobile node. Consequently, simultaneous communications between many source destination pairs are possible, and the spectrum efficiency can be increased.

Second, the adaptive beam and null controls, which maximize the SINR value depending on each specific radio propagation situation, improve link quality significantly. Moreover, intelligent digital signal processing algorithms can constructively combine the line of sight and delayed signals from the desired user and make link quality more stable and reliable.

Third, the antenna array forms the directive beam by combining the signals from all branches, and the link distance would be increased according to the number of branches. Consequently, the required number of hops from source to destination can be reduced, compared to omni-directional use. Moreover, the transmission power control relaxes the transmission power and realizes lower power consumption.

Thus, SDMA (Space Division Multiple Access) by a smart antenna in a wireless ad hoc network realizes not only throughput improvement but also less delay and longer battery life, simultaneous connections within omni range, and a larger number of link candidates for multi-hop routing to improve reliability and scalability.

3.2.2 Classification of Smart Antennas

There are two types of smart antennas used in the context of wireless networks: switched-beam or fixed-beam antennas and steerable adaptive array antennas.[17,18] A switched-beam antenna generates multiple predefined, fixed directional beam patterns and applies one at a time when receiving a signal. It is the simplest technique and consists of only a basic switching function between separate directive antennas or predefined beams of an array of N antenna elements that are deployed into non-overlapping fixed sectors, each spanning an angle of $360/N$ degrees. Signals will be sensed in all sectors and the antenna is capable of recognizing the sector with the maximum gain. When receiving, exactly one sector, which usually is the one chosen by the sensing process, will collect the signals.

On the other hand, the beam structure of a steerable adaptive array antenna, which is more advanced than a switched-beam antenna, adapts to the radio frequency (RF) signal environment and directs beams toward the signal of interest to maximize the antenna gain, simultaneously depressing the antenna pattern (by setting nulls) in the direction of the interferers.[17] In adaptive array antennas, an algorithm is needed to control the output, that is, to maximize the SINR.

The difference between both kinds of smart antennas is as follows: fixed-beam antennas focus their smartness in the strongest signal beam detection and adaptive array antennas benefit from all the received information within all antenna elements to optimize the output SINR through a weight vector adjustment.

3.2.3 ESPAR: A Smart Antenna for Wireless Ad Hoc Networks

Smart antennas have many features, such as capacity increase, range extension, better signal quality, etc., and can be deployed at the base stations of cellular networks. However, there are many issues to implement a smart antenna at mobile terminals for wireless ad hoc networks. A smart antenna requires an individual RF circuit for each element of antenna arrays. This factor increases the cost, power, and size for array control, such as optimization and tracking. The ESPAR (Electronically Steerable Parasitic Array Radiator) antenna addresses this drawback.[14–16]

Most smart antennas use DBF (digital beamforming) as the adaptive beamforming architecture. The received signal at each antenna branch is changed to the baseband by a down converter, and then this down-converted digital signal is fed into an AD converter. Finally, the digital signal is used as the input parameters of DSP, which optimize the beam shape adequately. The weight control of adaptive beamforming is performed in the digital part of each branch. Because the frequency down converter, AD converter, and the weight control function are required for each branch, DBF is not suitable for terminal use. Figure 3.2(a) shows DBF architecture. On the other hand, the ESPAR antenna takes the analog approach, or ABF (aerial beamforming) architecture. ABF is RF beamforming that reactively controls antenna arrays and drastically reduces the circuit complexity. Figure 3.2(b) shows ABF architecture. The ESPAR antenna consists of one center element connected to the source (the main radiator) and several surrounded parasitic elements (typically four to six passive radiators) in a circle. Figure 3.2(c) shows the configuration of a seven-element ESPAR antenna. Each parasitic element (the passive radiators) will be reactively terminated to ground. Adjusting the value of the reactance that terminates the parasitic elements forms the antenna array radiation pattern into different shapes. Because the beam is controlled aerially in the RF stage through the mutual coupling between center element and the surrounding parasitic elements, the weight control part of the ESPAR antenna is very simple, compared to DBF architecture. Also, ESPAR uses only one frequency down converter and one AD converter. Thus, ESPAR antennas can realize low-cost, smaller-size, and lower-power consumption, and they are suitable for terminal applications.

The ESPAR antenna can also be used as a generalized switched-beam antenna or quasi-switched-beam antenna by selecting the value of reactance for one specific directional beam among multiple directional beam patterns, without using multiple receiver chains (frequency converters and analog-digital converters). By including some mechanism to detect direction of arrival (DoA) for the signal received from the user, continuous tracking can be achieved and it can be viewed as a generalization of the switched-beam concept. In this case also, the received power is maximized. The advantage of using an ESPAR antenna as a generalized switched-beam antenna is that, with only one receiver chain, continuous tracking is possible and we can have a variable number of beam patterns. In other words, the directional beams that are formed with an ESPAR antenna when used in switched-beam mode need not be restricted to non-overlapping fixed sectors, each spanning an angle of $360/N$ degrees, as in the case of conventional switched-beam antennas with N elements.

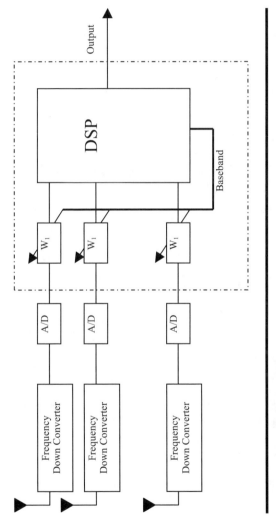

Figure 3.2(a) DBF architecture.

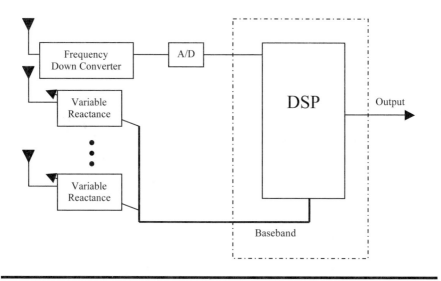

Figure 3.2(b) ABF architecture.

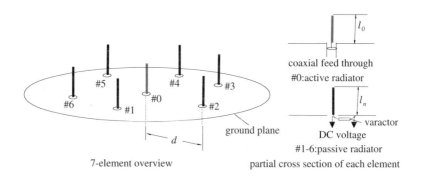

Figure 3.2(c) Configuration of the ESPAR antenna.

When the received power is maximized under the assumption that each component of the voltage is equal, the derived voltage vector creates an omni-directional beam pattern. On the other hand, when the received signal power is maximized in the direction of the target terminal, a sector-beam pattern is formed. Depending on the direction of the desired beam, ESPAR forms a sector-beam pattern differently (Figure 3.3). Figure 3.3(a) shows the pattern at 0 degrees: a beam pattern formed at each antenna element at an interval of 0 to 60 degrees, 60 to 120 degrees, and so on, thus forming six beams. Figure 3.3(b) shows the pattern at 30 degrees: a beam pattern formed at each

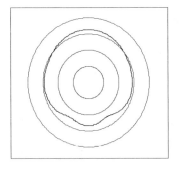 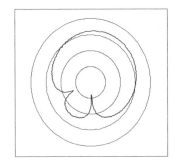

(a) ESPAR pattern at 0 degrees   (b) ESPAR pattern at 30 degrees

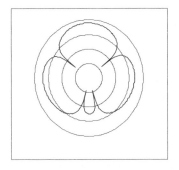 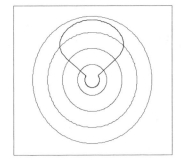

(c) DEFAULT switched beam   (d) IDEAL directional antenna
antenna pattern in QualNet

Figure 3.3 Different directional antenna patterns used for simulation.

in-between antenna elements at an interval of 30 to 90 degrees, 90 to 150 degrees, and so on, thus forming six more patterns. Together they constitute 12 overlapping patterns at 30-degree intervals. Figure 3.3(c) shows the QualNet default antenna pattern with a 45-degree beamwidth and Figure 3.3(d) shows an ideal directional antenna with a 45-degree beamwidth and insignificant side lobes.

As will be demonstrated in performance evaluation, the performance of an ideal directional antenna is the best (as expected); at the same time, ESPAR performance is much better than the default antenna pattern of QualNet. The reason for this is that the pattern of ESPAR has less coverage area with less pronounced side lobes, as compared to the default antenna pattern in QualNet.

As a steerable beam and null adaptive array antenna, the ESPAR antenna tries to maximize the output SINR under the desired and interfering signals. To distinguish the original desired signal from the combined interfering multiple signals, we can use a training sequence that is included within the transmitted packet.[20] DBF can estimate the difference by the MMSE (minimum mean square error) criterion. However, ESPAR is based on analog beamforming, and it is very difficult to estimate the nonlinear relation between the reactance and the weight. Consequently, ESPAR controls the value of the reactance by the MCCC (maximum cross correlation coefficient) criterion. Moreover, even if there is no information about the transmitted signal, ESPAR can form an adequate beam pattern, for example, through CMA (constant modulus algorithm).[21]

ATR-ACR has developed a wireless ad hoc network testbed using ESPAR to experimentally assess the performance of MAC and routing protocols when directional and adaptive beam patterns are used for data communication. This testbed consists of a smart antenna, wireless module, and notebook PC with PCMCIA card. Beam control and directional MAC functions are implemented in the wireless module. Directional MAC function not only handles the four-way handshake with the instruction of beam pattern to beam control module, but also periodically maintains the directions and signal levels of surrounding terminals as the AST in order to realize SDMA. Some experimental results of directional MAC with AST are shown in Ueda et al.[13]

3.2.4 Issues of Smart Antennas in the Context of Wireless Ad Hoc Networks

More efficient processing algorithms are required for faster convergence with less overhead, which is suitable for ad hoc networks. From the radio environment point of view, the various kinds of impairments (fast fades Doppler spread, slow fades delay spread, angle spread, co-channel interference) should be considered, and the applicable number of interferences, limit of angle separation between them and null and the performance under indoor/pedestrian/vehicular conditions should be clarified. These results are strongly related to the current MAC and routing protocols, which should be customized for the utilization of smart antennas.

Second, the advantages of an adaptive array antenna should be shown under the conditions of mobility and different node densities with the utilization of the location-tracking mechanism, such as MAC-based[22] and MUSIC.[23]

Third, power control in conjunction with adaptive beamforming and null steering is the issue to be investigated, which also impacts MAC and routing protocols.

3.3 Location-Tracking Mechanisms for Neighborhood Discovery

As indicated earlier, developing a suitable MAC protocol in an ad hoc network to exploit the advantages of a directional antenna for overall performance improvement requires proper location tracking and neighborhood knowledge. The source and destination nodes identify each other's direction during omni-directional RTS/CTS exchange.[1] However, in this mechanism, a node is not aware of its complete neighborhood information. In some studies, the use of GPS is proposed to track the location of each node but the exact mechanism of information exchange and the consequent overhead has not been discussed.[7,11] Bandyopadhyay et al.[8,9] have developed a MAC protocol, where each node keeps certain neighborhood information dynamically through the maintenance of an angle signal table. In this method, to form the AST, each node periodically sends a directional beacon in the form of a directional broadcast, sequentially in all directions at 30-degree intervals, covering the entire 360-degree space. The nodes, which receive these signals at different angles, determine the best-received signal strength and transmit the information back to the source node as a data packet with an RTS/CTS handshake. However, the overhead due to control packets is very high in this method[9] of location tracking. In this section, we will illustrate a receiver-oriented location-tracking mechanism to reduce the control overhead.

In the proposed protocol, each node waits in omni-directional-receive mode while idle. Whenever it senses some signal above a threshold, it enters into rotational-sector-receive mode. In rotational-sector-receive mode, node n rotates its directional antenna sequentially in all directions at 30-degree intervals, covering the entire 360-degree space in the form of the sequential directional receiving in each direction and senses the received signal at each direction. After one full rotation, it decides the best possible direction of receiving the signal with maximum received signal strength. Then it sets its beam to that direction and receives the signal.

However, to enable the receiver decoding of the received signal, each control packet is transmitted with a preceding tone with a duration such that the time to rotate a receiver's rotational receive beam through

360 degrees is less than the duration of the tone. The purpose of this transmitted tone before any control packet is to enable the receiver to track the best possible direction of receiving the signal. Once it sets its beam to that direction, the purpose of the tone signal is over and subsequently the control packet is transmitted.

In this proposed framework, we have used three types of control packets: beacon or hello packets used to track the location of neighboring nodes, RTS (request to send) packets, and CTS (clear to send) packets. Each beacon is a periodic signal, transmitted from each node at a predefined interval. At each periodic interval, each node, say m, sends an omni-directional beacon to its neighbors if the medium is free. As indicated earlier, each beacon is transmitted with a preceding tone signal that helps the receivers to detect the best possible direction of receiving the beacon. Then each receiver sets its beam to that direction and receives and decodes the beacon. Thus, the node n, which is, say, a neighbor of m, forms the angle-signal information for node m, and similarly, for other neighbors. An entry in AST of node n for its neighbor m is $SIGNAL_{n,m}(t)$, which is the maximum strength of the received signal at node n from node m at an angle with respect to n and as perceived by n at any point of time t. Based on AST, a node n knows the direction of node m and controls the medium access during transmission-reception.

Because RTS is a broadcast packet and contains the source address, nodes can decode that RTS also to form the AST. So, we have used RTS as a beacon. If an RTS is sent, the beacon timer is reset. The use of RTS as a beacon is advantageous at high traffic where overhead due to a beacon is minimized. This is because the transmitting nodes do not have to send an additional beacon to inform its neighbors of its presence.

3.4 Directional Media Access Control Protocols

In IEEE 802.11 MAC protocol standard, the RTS-CTS-DATA-ACK exchange mechanism is used to ensure reliable data communication. In our scheme, initially, when node n wants to communicate with m, it senses the medium and if it is free, sends an omni-directional RTS. The back-off mechanism is the same as in IEEE 802.11. The purpose of the RTS is to inform all the neighbors of n, including m, that a communication from n to m has been requested. It also specifies the approximate duration of communication. All the neighboring nodes of n keep track of this request from node n, whose direction is known to each of them from the received RTS signal. The mechanism for receiving the RTS is the same as that for a beacon.

The target node m sends an omni-directional CTS to grant the request and to inform the neighbors of m that m is receiving data from n. It also specifies the approximate duration of communication. All the neighboring nodes of m keep track of the receiving node m, whose direction is known to each of them from the received CTS signal. Once again, the mechanism for receiving the CTS is the same as that for a beacon.

Several MAC protocols using directional antennas have been proposed, as discussed in Chapter 2. In this work, we have used omni-directional RTS/CTS and directional data/acknowledgment. It is to be noted that the objective of RTS/CTS here is not to inhibit the neighbors of n and m from transmitting or receiving (as is the case with an omni-directional antenna) but to inform the neighbors of n and m that m is receiving data from n. Moreover, our rotational-sector–based mechanism provides a unified framework for location tracking and MAC with minimal overhead.

After transmission of an omni-directional CTS, the receiving node waits in directional-receive mode until data is transmitted or times out and returns to omni-directional-sensing mode. Also, once the CTS is received, the transmitter transmits data directionally and waits for acknowledgment directionally until acknowledgment is received or times out and returns to omni-directional-sensing mode. The directional reception mode ensures proper reception of the signal from the required direction and minimization of interference from the other direction.

Other nodes in the neighborhood of n and m, who overheard the RTS/CTS exchange, set their directional network allocation vector (DNAV) in the direction that they detected as the direction of arrival of the RTS or CTS, respectively. Now, if they have a packet to send to a node, whose direction (as known from the AST) is not in the direction of the blocked DNAV, then they can issue both RTS and CTS omni-directionally without disturbing the communication between n and m. If the direction of receiving node is blocked by DNAV and an RTS is issued, it is most probable that a CTS will not be issued or there may be an RTS collision. As a result, the node will increase its contention window and enter into back-off mode. This may happen repeatedly and, as a result, the node will get less chance to transmit. So we do not allow RTS transmission in this case. Here, the node waits for DNAV time and then tries to start communication, which is similar to waiting for NAV as explained in standard IEEE 802.11.

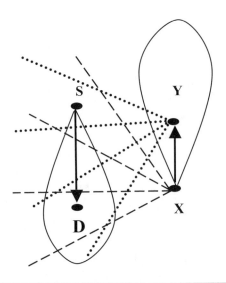

Figure 3.4 Multiple simultaneous communications.

Figure 3.4 illustrates the mechanism of two simultaneous communications in the same region. Let us assume that nodes S and D are communicating. The directional beam from S covering D is shown in the figure. Now, another pair of nodes, X and Y, both in the omni-directional neighborhood of S and D, desires to communicate. Both of them have already received RTS/CTS from S–D. From their respective ASTs, X knows the angular position of S and D with respect to X, and Y also knows the angular position of S and D with respect to Y. Both X and Y will set the DNAV toward S and D. If the directional beam from X to Y captures S or D, then the node X has to sit idle until time-out mentioned in DNAV and thus X has to defer its desire. Otherwise, node X can issue an RTS. In other words, a node can issue an RTS only if this communication does not intrude into the area of existing communications. Figure 3.5 shows the state transition diagram of the proposed scheme.

3.5 A Few Assumptions and the Rationales

When the antenna of a node is operating in omni-directional mode, it is capable of transmitting and receiving signals from all directions with a gain, say, G^{omni}. While idle, a node operates in omni-directional-receive mode.

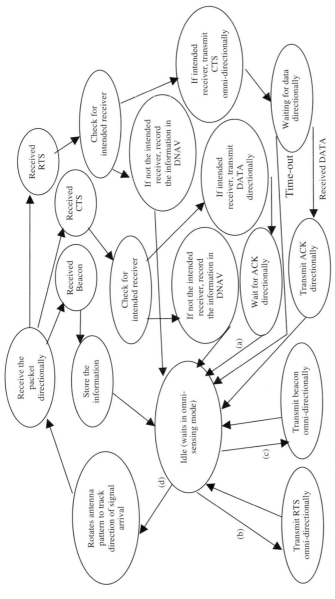

Figure 3.5 State transition diagram of the proposed directional MAC scheme.

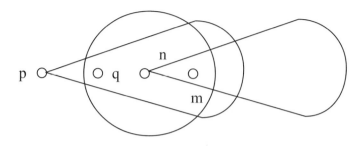

Figure 3.6 The capture of receiver m by transmitter n is strong enough to tolerate interference from another transmitter p.

When the antenna of a node operates in directional mode, a node can point its beam (main lobe) toward a specified direction with beamwidth w and with a gain, say G^{dir} ($G^{dir} \gg G^{omni}$). Beamwidth is around 60 degrees in our simulation.

Consequently, for a given amount of input power, the transmission range R^{dir} with a directional antenna will be much larger than that with a corresponding omni-directional antenna (R^{omni}).

We define neighbors of a node n as a set of nodes within the omni-directional transmission range of n. They are assumed to be one hop away from n. It implies that a node outside the omni-directional transmission range of n will not be considered as a neighbor of n, even if it is reachable by n in one hop using a directional beam from n formed toward that node. From the perspective of directional data communication, it implies that a neighbor, say m of a node n, is always a strong neighbor. As shown in Figure 3.1, when a node n forms a directional beam toward its neighbor m, m is well within the transmission zone so formed. Hence, the received signal strength at m from n is always high to ensure proper capture even in the presence of other interferences. Thus, the chance of m getting disconnected or weakly connected during a data-packet transfer from n due to an outward mobility of either m or n is far less.

This will alleviate the problem of a hidden terminal in this context.[7] Let us consider Figure 3.6, where node n is communicating with node m with a directional beam. Node p now wants to communicate with node q. If node p is within the neighborhood of n, this communication will not be initiated, because p is not allowed to form a directional beam toward n or m. However, if node p is outside the neighborhood of n, node p forms a directional beam toward node q and starts communication. This may interfere with node m's reception. However, because the distance between p and m is larger than between n and m by at least R^{omni} (the omni-directional range), the received signal at m from n will predominate and the chance of data packets being lost due to this interference will be insignificant.

However, as a consequence of this assumption, we are sacrificing multi-hop efficiency, which could have been achieved using a directional antenna, because with a larger range of directional beam a destination is reachable in a fewer number of hops as compared with an omni-directional antenna. However, what we are gaining is SDMA efficiency, as will be demonstrated in the performance evaluation.

3.6 Performance Evaluation

3.6.1 Simulation Environment

The simulations are conducted using QualNet 3.1.[12] We simulated an ESPAR antenna in the form of a quasi-switched-beam antenna, which is steered discretely at an angle of 30 degrees, covering a span of 360 degrees. We have simulated our MAC protocol with (1) a simulated ESPAR antenna pattern (ESPAR), (2) QualNet's default directional antenna pattern, and (3) an ideal directional antenna pattern without side lobes as described in Section 3.2. We have done the necessary changes in the QualNet simulator to implement directional virtual carrier sensing in MAC layer and directional transmission in physical layer. In our simulation, we have chosen the duration of the preceding tone in control packets to be 200 microseconds (ms), based on the hardware performance of ESPAR.

We used simple, one-hop, randomly chosen communications to avoid the effects of routing protocols to clearly illustrate the difference between 802.11 and our proposed MAC. Also, we used static routes to stop all the packets generated by any routing protocol, whether it is proactive or reactive. In our simulation, we studied the performance of the proposed MAC protocol in comparison with the existing omni-directional 802.11 MAC protocol by varying the data rate and number of simultaneous communications. In studying our MAC protocol, we used different antenna patterns as described above to ensure the robustness of our proposed MAC protocol. In doing this, we used ESPAR as one of the antenna patterns, to evaluate the performance of the ESPAR antenna as well.

Forty nodes were randomly placed over a 1000 × 1000 square meter area. The simulation was conducted in two steps. First, keeping the number of simultaneous communications constant at ten, the data rate

Table 3.1 Parameters Used in Simulation

Parameters	Value
Area	1000 x 1000 sq. m
Number of nodes	40
Transmission power	15 dBm
Receiving threshold	−81.0 dBm
Sensing threshold	−91.0 dBm
Data rate	2 Mbps
Packet size	512 bytes
Duration of preceding tone in RTS/CTS/Beacon	200 ms
CBR packet arrival interval	2 ms to 50 ms
Number of simultaneous communications	4 to 12
Simulation time	5 minutes

was gradually increased from 81.92 Kbps (512 bytes of data packets injected at an interval of 50 ms) to 2.048 Mbps (512 bytes of data packets injected at an interval of 2 ms). Second, in keeping the data rate constant at 409.6 Kbps (512 bytes of data packets injected at an interval of 10 ms), the number of simultaneous communications increased from four to twelve. In both steps, we evaluated average throughput and one-hop average end-to-end delay.

The set of parameters used is listed in Table 3.1.

3.6.2 Results

We used the existing IEEE 802.11 MAC, which we caption "802.11," as a benchmark to compare and evaluate the performance of our proposed MAC protocol with ESPAR, QualNet's default antenna, and an ideal antenna, respectively. Our evaluation is based on two criteria: average throughput and one-hop average end-to-end delay.

The results are shown in Figures 3.7 and 3.8, respectively. Each result reported is an average of ten executions with different seeds. So, to complete our results, we had to simulate over 400 scenarios, each of which was executed in the simulator for 5 minutes to get an overall average result.

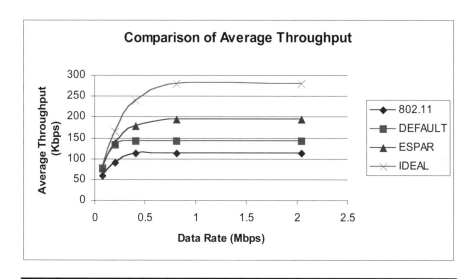

Figure 3.7(a) Comparison of average throughput of IEEE 802.11 and E-MAC with an ESPAR antenna, a QualNet default antenna, and an ideal antenna with increasing data rate at ten simultaneous communications.

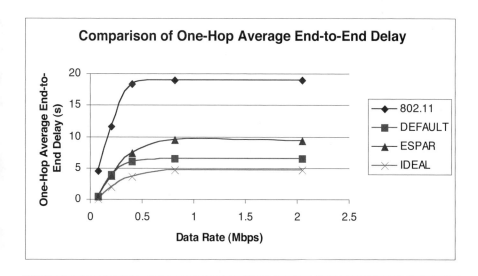

Figure 3.7(b) Comparison of one-hop average end-to-end delay of IEEE 802.11 and E-MAC with an ESPAR antenna, a QualNet default antenna, and an ideal antenna with increasing data rate at ten simultaneous communications.

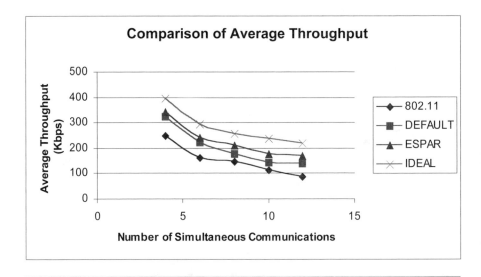

Figure 3.8(a) Comparison of average throughput of IEEE 802.11 and E-MAC with an ESPAR antenna, a QualNet default antenna, and an ideal antenna with an increasing number of simultaneous communications at a constant data rate of 100 packets per second where each packet is 512 bytes.

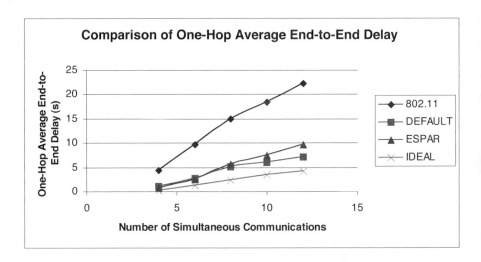

Figure 3.8(b) Comparison of one-hop average end-to-end delay of IEEE 802.11 and E-MAC with an ESPAR antenna, a QualNet default antenna, and an ideal antenna with an increasing number of simultaneous communications at a data rate of 100 packets per second where each packet is 512 bytes.

In Figure 3.7(a), it is seen that with increasing data rate, average throughput of our proposed MAC protocol (E-MAC) with any directional antenna pattern is much better than that of IEEE 802.11 with an omni-directional antenna. It is also seen in Figure 3.7(b), that one-hop average end-to-end delay performance of E-MAC with any directional antenna is much better than that obtained with IEEE 802.11 protocol.

In omni-directional 802.11, nodes have to enter in a back-off state more often as they find the medium busy. With an increasing data rate, contention in MAC increases. But with the use of a directional antenna, and the implementation of directional virtual carrier sensing, E-MAC creates an environment of lower contention that "802.11" cannot create with an omni-directional antenna. Hence, with an increasing data rate, average throughput increases sharply in E-MAC as shown in Figure 3.7(a). In E-MAC, once RTS/CTS handshaking is done, a node transmits and receives data and acknowledgment directionally with high gain. So, the chance of missing the data at the receiver end or the acknowledgment at the transmitter end is minimized. But, in 802.11, the chance of missing data is even more than that of RTS/CTS. This is due to two reasons: (1) data is sent with the same gain as in RTS/CTS omni-directionally and received omni-directionally, and (2) data is a large packet compared to RTS/CTS and proper reception requires the SINR level to remain high for a longer period of time. These reasons also account for higher average throughput and lower end-to-end delay in E-MAC as compared to 802.11.

MAC performance depends much on the directional antenna pattern also. So, we have simulated for three different types of directional antenna patterns. The QualNet default is a standard antenna pattern. The ideal antenna is an ideal directional antenna pattern with no side lobes. This is already illustrated in Section 3.2. Average throughput with ESPAR is better than the QualNet default antenna, and the gain in throughput obtained with ESPAR is nearly 1.8 times that of IEEE 802.11.

In Figure 3.8(a), it is observed that with the increasing number of simultaneous communications, average throughput decreases in both E-MAC and 802.11, but E-MAC shows significant gain in average throughput. This is because E-MAC does not inhibit neighboring nodes to transmit but just informs neighbors of the ongoing communication and its direction, so that they can start communication in other directions. But 802.11 with an omni-directional antenna keeps all neighboring nodes silent by issuing RTS/CTS. Also, with the increasing number of simultaneous communications, average end-to-end delay (one-hop) increases in both IEEE 802.11 and E-MAC, as shown in

Figure 3.8(b), but the increase is much more prominent in 802.11 than in E-MAC, irrespective of the directional antenna pattern used. With the increasing number of simultaneous communications, interference to each communication increases due to the interference of added simultaneous communications. But E-MAC not only informs other nodes in its vicinity of the ongoing communication, but also transmits and receives directionally with larger capture, which minimizes interference from other directions also. Thus E-MAC exploits SDMA efficiency for which more simultaneous communications are possible, which leads to lesser queuing delay and lesser one-hop average end-to-end delay as observed in Figure 3.8(b).

3.7 Discussion

The use of directional antennas in an ad hoc wireless network can drastically improve system performance, if a proper MAC protocol can be designed. With directional setting of virtual carrier sensing, the medium can be utilized to its maximum with directional antennas. With a minimum overhead of location tracking, gain obtained in MAC is really significant. Also, the success of the MAC protocol highly depends on the directional antenna pattern. Average throughput with ESPAR is better than the QualNet default antenna, and the gain in throughput obtained with ESPAR is nearly 1.8 times than that of IEEE 802.11. The location-tracking mechanism as done in our proposed MAC protocol can be utilized in designing efficient routing protocol also, as will be discussed in the next chapter.

References

1. Nasipuri, A. et al., A MAC Protocol for Mobile Ad Hoc Networks Using Directional Antennas, in *Proc. IEEE WCNC*, 2000.
2. Nasipuri, A., Li, K., and Sappidi, U.R., Power Consumption and Throughput in Mobile Ad Hoc Networks Using Directional Antennas, in *Proc. IEEE International Conference on Computer Communication and Networks*, Miami, Florida, October 14–16, 2002.
3. Zander, J., Slotted ALOHA Multihop Packet Radio Networks with Directional Antennas, in *Electronic Letters*, Vol. 26, No. 25, 1990.
4. Kobayashi, K. and Nakagawa, M., Spatially Divided Channel Scheme Using Sectored Antennas for CSMA/CA–Directional CSMA/CA, in *Proc. PIMRC*, 2000.

5. Takai, M. et al., Directional Virtual Carrier Sensing for Directional Antennas in Mobile Ad Hoc Networks, in *Proc. ACM MobiHoc*, June 2002.
6. Ramanathan, R., On the Performance of Ad Hoc Networks with Beamforming Antennas, in *Proc. ACM MobiHoc*, October 2001.
7. Roy Choudhury, R. et al., Using Directional Antennas for Medium Access Control in Ad Hoc Networks, in *Proc. ACM MOBICOM*, Atlanta, Georgia, September 2002.
8. Bandyopadhyay, S. et al., An Adaptive MAC and Directional Routing Protocol for Ad Hoc Wireless Network Using Directional ESPAR Antenna, in *Proc. ACM Symposium on Mobile Ad Hoc Networking & Computing*, Long Beach, California, October 4–5, 2001.
9. Bandyopadhyay, S. et al., An Adaptive MAC Protocol for Wireless Ad Hoc Community Network (WACNet) Using Electronically Steerable Passive Array Radiator Antenna, in *Proc. GLOBECOM*, San Antonio, Texas, November 25–29, 2001.
10. Yum, T.S. and Hung, K.W., Design Algorithms for Multihop Packet Radio Networks with Multiple Directional Antennas Stations, in *IEEE Transactions on Communications*, Vol. 40, No. 11, 1992, p. 1716.
11. Ko, Y.B., Shankarkumar, V., and Vaidya, N.H., Medium Access Control Protocols Using Directional Antennas in Ad Hoc Networks, in *Proc. IEEE INFOCOM*, March 2000.
12. QualNet Simulator Version 3.1, Scalable Network Technologies, www.scalable-networks.com.
13. Ueda, T. et al., Evaluating the Performance of Wireless Ad Hoc Network Testbed with Smart Antenna, in *Proc. Fourth IEEE Conference on Mobile and Wireless Communication Networks (MWCN2002)*, September 2002.
14. Ohira, T., Adaptive Array Antenna Beamforming Architectures as Viewed by a Microwave Circuit Designer, in *Proc. Asia-Pasific Microwave Conf.*, Sydney, December 2000.
15. Ohira, T. and Gyoda, K., Electronically Steerable Passive Array Radiator (ESPAR) Antennas for Low-cost Adaptive Beam Forming, in *Proc. IEEE International Conference on Phased Array Systems*, Dana Point, California, May 2000.
16. Gyoda, K. and Ohira, T., Beam and Null Steering Capability of ESPAR Antennas, in *Proc. IEEE AP-S International Symposium*, July 2000.
17. Lehne, P.H. and Pettersen, M., An Overview of Smart Antenna Technology for Mobile Communications Systems, in *Proc. IEEE Communications Surveys*, Vol. 2, No. 4, 1999.
18. Liberti, J.C. and Rappaport, T.S., *Smart Antennas for Wireless Communications: IS-95 and Third Generation CDMA Applications*, Prentice-Hall, Upper Saddle River, New Jersey, 1999.
19. Cheng, J. et al., Electronically Steerable Parasitic Array Radiator Antenna for Omni- and Sector-pattern Forming Applications to Wireless Ad Hoc Networks, in *Proc. Inst. Elect. Eng. Antennas and Propagation*, Vol. 150, August 2003, p. 203.

20. Ohira, T., Analog Smart Antennas: An Overview (invited), in *Proc. IEEE Int'l Symp. Personal Indoor Mobile Radio Commun., PIMRC2002*, Vol. 4, Lisbon, Portugal, September 2002, p. 1502.
21. Ohira, T., Blind Aerial Beamforming Based on a Higher-order Maximum Moment Criterion (Part I: Theory), in *Proc. Asia-Pacific Microwave Conference*, WE3C-1, Kyoto, Japan, November 2002, p. 181.
22. Ueda, T., Bandyopadhyay, S., and Hasuike, K., System Performance of Adaptive MAC Protocol for SDMA through Overhead Reduction in Wireless Ad Hoc Networks, in *Transactions of The Institute of Electronics, Information and Communication Engineerings (IEICE)*, Vol. J85-B, No. 12, December 2002 (special issue for ad hoc networks).
23. Taillefer, E. et al., Reactance-Domain MUSIC for ESPAR Antenna, in *Proc. WCNC*, 2003.

Chapter 4

Location Tracking and Location Estimation of Nodes in Ad Hoc Networks: A Testbed Implementation

4.1 Introduction

In this chapter, we will explore another advantage of directional antennas in estimating approximate location of nodes without using any additional hardware like GPS (global positioning system). We set up a testbed of ad hoc networks using directional antennas and demonstrated the effectiveness of directional tracking of the neighborhood of each node and subsequently location estimation of each node with respect to two reference nodes. This neighborhood tracking and location estimation not only help us in implementing directional MAC and directional routing protocols like location-aided routing, but also help us in applications involving location-based services where each node needs to know the approximate locations of other nodes in the network.

As indicated in Chapter 3, to fully exploit the capability of a directional antenna it is necessary for each node to know the information of the neighboring nodes (such as node-ID, direction, link quality, etc.) beforehand. Whenever a source and a destination node are engaged in a communication, all the neighbors of source and destination nodes should know the direction of communication so that they can initiate new communication in other directions, thus preventing interference with ongoing data communication between the source and destination. In other words, to implement effective MAC and routing protocol in this context, a node should know how to set its transmission direction to transmit a packet to its neighbors.[1] It becomes imperative to have a mechanism at each node to track the locations of its neighbors. Moreover, our testbed experience indicates that accurate location tracking and location estimation based on single beaconing is not possible due to fluctuating channel characteristics. Usually, the simulation-based studies do not address this issue.[2] In this chapter, we will show that the use of multiple beacons and multiple observations on location data are required for location tracking and location estimation. This limits the applicability of our mechanism in highly mobile scenarios. However, in semi-static scenarios (where mobility is infrequent or only a few nodes are mobile), this scheme works well. Areas of application include sensor networks for remote environment monitoring (forest, mine, natural disaster-prone area, etc.) or community networks in remote areas (e.g., geographic exploration or hiking and trekking in a remote area).

Several location-estimation techniques have been proposed by contemporary researchers that do not rely on GPS. Such GPS-free positioning systems are especially suitable for indoor network environments. In general, a location-estimation algorithm for a distributed environment such as an ad hoc network needs to handle some specific design issues. The algorithm should be less complex, be cost-effective, and should provide fairly accurate location estimation so that the corresponding relative positions of the nodes match the actual network topology.

Existing location-discovery techniques typically use distance or angle measurements from a fixed set of reference points and apply lateration or triangulation to solve for the unknown location.[3,4] The distance or angle estimates may be obtained from received signal strength (RSSI) measurements; time-of-arrival, time-difference-of-arrival measurements (ToA, TDoA); and angle-of-arrival (AoA) measurements. It was argued that non-uniform propagation environments make RSSI methods unreliable and due to the high propagation speed of wireless signals, a small error can cause a large error in the distance estimate in the case of ToA, TDoA. So, the localization techniques using ToA, TDoA need to use a signal with a slower propagation speed than wireless such as ultrasound and this in turn can give an accurate result but requires additional hardware to receive the ultrasound signal. A combination of RSSI and other measurements are suggested for a reliable estimation.[4,5] The RADAR system[4] is designed for indoor localization, which uses extensive RF signal strength measurements that are performed offline to design signal strength maps. These maps are used for localization to measure the distance from signal strength measurements. A ToA location-sensing system includes GPS and the Active Bat Location System.[6] The BAT system[7] uses an array of ultrasound receivers for processing received signals from a user of unknown location. In the Cricket location support system,[8] fixed beacons broadcast local geographical information to the listener nodes to increase accuracy of distance estimation from ultrasound signals. An alternative way uses the measurements of the attenuation of a signal, which is essentially the decrease in signal strength relative to the original intensity. The SpotON ad hoc location system[9] and Active Campus[10] implements signal attenuation–based measurement instead of distance measurements to calculate the position of an object.

Capkun et al. have proposed a distributed infrastructure-free positioning algorithm for the nodes in ad hoc networks that does not rely on GPS.[11] Their algorithm uses the distance between nodes to build a relative coordinate system in which node positions are computed in two dimensions. They have indicated that this kind of algorithm can be used for applications like location-aided routing and geodesic packet forwarding. They have assumed that all the nodes have the same technical characteristics, links between nodes are bidirectional, nodes use omni-directional antennas, and the mobility is restricted. Each node builds its local coordinate system assuming that the node itself is the center (0, 0) of its own coordinate system. It will then compute the positions of its neighbors according to its local coordinate system using a distance-based triangulation technique. Capkun et al. also described how to adjust the direction of the local coordinate system of the nodes to obtain the same direction for all the nodes in the network. However, it has been shown that location estimation using angle (angulation) performs better than location estimation using distance (lateration), because angle measurement noise is much smaller than distance measurement noise based on signal strength.[4]

Nasipuri and Lee have proposed a location-estimation technique for sensor networks based on the angle of arrival of beacons from three or more fixed beacon nodes whose positions are known.[3] This method is based on triangulation technique and thus does not involve much complexity and cost. So, it is suitable for an ad hoc sensor network environment. A major point raised by the authors is that an ideal location-estimation algorithm should be capable of determining the location of a node, which is not within the direct transmission range of the fixed beacon nodes. Lee et. al also supported the fact that the location-estimation protocols should have the capability of multi-hop-based location computing and they proposed a location-sensing protocol utilizing the directional antenna system and the direction of arrival (DoA) estimation algorithm.[4] They observed that the location-sensing protocol utilizing the directional antenna system generates more accurate results than the similar protocols using the omni-directional antenna, especially distance-based triangulation technique. They have assumed a switched-beam antenna system with reference nodes and either a switched-beam or omni-directional antenna system with mobile nodes. However, their proposal has not been validated with a real directional antenna. Moreover, they have not addressed the issue of coordinate synchronization of intermediate nodes, acting as secondary reference nodes in case of multi-hop location estimation.

In this chapter, our objective is to actually implement an angle-of-arrival (AoA)-based location-estimation technique on an ad hoc network testbed using a directional antenna and to evaluate the effectiveness of the proposed methodology in a real-world environment. This is probably the first such implementation with a directional antenna that attempts to evaluate the effectiveness of location tracking and multi-hop location estimation of nodes in ad hoc networks.

4.2 Location Tracking and Neighborhood Discovery

4.2.1 Formation of the NLST (Neighborhood Link State Table)

To effectively communicate with a neighbor using a directional antenna, a node needs to know the exact direction of each of its neighbors to set its antenna beam to talk to that neighbor. Each node does this directional location tracking periodically. Normally, each node waits in omni-directional-receive mode while idle. To initiate location tracking, a node, say n, broadcasts 12 directional beacons sequentially,

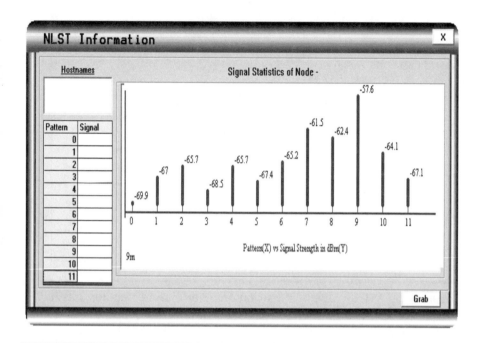

Figure 4.1 NLST at node i.

using 12 directional beam patterns of the ESPAR antenna. Each beacon contains node ID and corresponding direction of transmission. This is done sequentially in all directions at 30-degree intervals, covering a span of 360 degrees. A node, say i, in the neighborhood of n, which was waiting in omni-directional-receive mode will receive each directional broadcast packet and will record the received signal strength as perceived by i in each direction (SIGNAL$^\alpha_{ni}$, $\alpha=\{0°,30°,60°,....,330°\}$) from node n to node i. In this way, a neighborhood link-state table (NLST) will be generated at i for node n. Figure 4.1 illustrates the typical structure of an NLST at a node i. Similarly, all nodes will issue periodic directional beacons and each node, say node i, will create NLST for all its neighbors in the format shown in Figure 4.1 and this is updated periodically.

4.2.2 Formation of the AST (Angle Signal Table)

After the formation of NLST, a node i will calculate the beam pattern of n at which it has received the strongest signal from n and the corresponding value of the received signal strength from n. Node i will then send this information to n in omni-directional mode. On

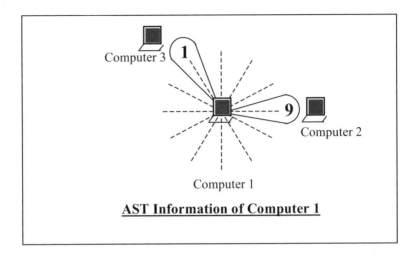

Figure 4.2 AST of computer 1 with computer 2 and computer 3 as its neighbors. Computer 1 can access 2 and 3 with beam patterns 9 and 1, respectively.

receiving that, node n will record this information in its angle signal table (AST). In a similar fashion, node n will receive similar information from all of its neighbors and update its AST. Thus, an AST at any node n essentially contains the neighbors of node n and the best possible direction to access each neighbor of n and the signal strengths as perceived by each neighbor in the corresponding best directions. This neighborhood information to construct AST is coming periodically at any node n (Figure 4.2). If node n does not receive directional information from any of its neighbors for a prespecified amount of time, the corresponding entry for that neighbor is deleted from its AST.

4.3 Location Estimation

4.3.1 Basic Idea

In this section we will discuss a simple mechanism to estimate the location of nodes in ad hoc networks. According to our proposal, a node will apply a simple triangulation technique based on the geometric properties of a triangle to compute its own location. Triangulation arguments refer to either distance (lateration) or to angle measurements (angulation). Lateration can be used to compute location from distance measurements from multiple reference positions, whereas angulation uses angle measurements with respect to multiple reference positions instead of the distance measurements to calculate the location of objects.[4,5] In two-dimensional space, two angle measurements and positions of two reference nodes are sufficient to uniquely identify the location of an object. So, our proposed location-estimation technique uses the AoA of the best signal to a node from a pair of reference nodes whose positions are fixed. The AoA measurements of best signals from neighboring reference nodes are available to a node in its NLST through location tracking/neighborhood discovery process. So, a node, which is within the range of a pair of reference nodes, will be able to calculate its own coordinate from the AoA of best signal from each reference node and the coordinates supplied by those two reference nodes. So, under the proposed framework it is possible to find the coordinate of a node that is within the range of the specified pair of reference nodes.

But, the problem arises if a node lies multi hops away from the reference nodes. In that case, it is not possible to estimate the location of that node using the above-mentioned scheme because the AoA measurements of signal from reference nodes are not available in such cases. So, we have suggested a multi-hop extension in our location-estimation mechanism in which the one-hop neighbors of a pair of reference nodes (primary reference nodes) will volunteer themselves as secondary reference nodes. As a result, a node, which is two hops away from primary reference nodes, will now use the secondary reference nodes to calculate its location per the above-mentioned procedure. Similarly, a three-hop neighbor of primary reference nodes will use the second-hop nodes as its reference nodes. Thus, it is possible to progressively estimate the location of all the nodes in the network.

The synchronization of the alignments of antenna of secondary reference nodes with the primary reference nodes is necessary to correctly determine the location of a two-hop node. Synchronization of alignment of an antenna with a reference node necessarily indicates the alignment of the 0th beam pattern of the antenna with that of the reference node. Otherwise, the frame of coordinate used by the one-hop neighbor of the primary reference point will be different from that used by two-hop neighbors. Synchronization of the orientation of antennas is essential to implement the common frame of reference for location estimation. A stepwise synchronization of antennas between primary and secondary reference nodes (and between secondary and tertiary reference nodes and so on) will make the common frame of reference for all the nodes in the network.

In our subsequent discussion, we will first explain the basic mechanism of calculation of the coordinate of a node with respect to a pair of primary reference positions. Next, we discuss elaborately the process of synchronization of antennas with reference nodes, which is a vital task to calculate the location of a distant node. Finally, we will illustrate the process of location estimation of a node that is multi hops away from primary reference nodes.

4.3.2 Location Estimation by a Node Using a Pair of Reference Nodes

First of all, two nodes are selected as two primary reference nodes manually and one is placed at (0, 0) and the other is placed at (x, 0) with their 0th beam patterns aligned across the X axis for the sake of simplicity. In our case, the coordinate of a node will be calculated on the basis of these two sets of reference coordinates and the corresponding angles of arrival of best signals to that node from those reference nodes. It is possible to get the information about the best beam pattern by which a neighbor can access a node from the NLST kept at each node. From the beam patterns it is possible to get the corresponding angles of arrival (e.g., angle of arrival of beam 0 is 0, beam 1 is 30, beam 2 is 60, and so on). So, the AoA of best signal to a node coming from reference nodes can be decoded from the corresponding beam patterns stored in the NLST of that node. We have used that in our coordinate calculation process.

Each reference node will periodically transmit its coordinate through directional beacons so that other nodes in the neighborhood of either of the reference nodes or both can collect that information for their coordinate calculation purpose.

If a node (say P in Figure 4.3) can manage to collect the coordinates of two reference nodes (that is possible only if it lies within the transmission range of both the reference nodes R1 and R2), then it will immediately calculate the coordinate of itself (p, q) using four parameters available to it: (1) coordinates of reference node R1 (x1, y1), which is (0, 0) in this case; (2) coordinates of R2 (x2, y2), which is (x, 0) in this case; (3) AoA of signal from R1 (α); and (4) AoA of signal from R2 (β). Using the process described below we can easily calculate the coordinates of P.

The general equation of line PR1 and PR2 will be:

$$(y - y1) = \tan \alpha \ (x - x1)$$

$$p = ((y2-y1) + x1 \tan \alpha - x2 \tan \beta)/(\tan \alpha - \tan \beta)$$
$$q = ((x1 - x2) \tan \alpha \tan \beta + (y2 \tan \alpha - y1 \tan \beta))/(\tan \alpha - \tan \beta) \quad (1)$$

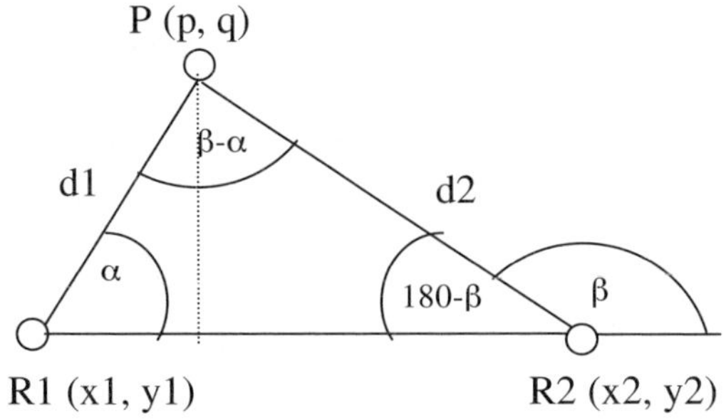

Figure 4.3 Illustration of coordinate calculation process of a node P using two given reference nodes R1 and R2.

$$(y - y2) = \tan \beta (x - x2)$$

Now, P (p, q) is the point of intersection of both PR1 and PR2. So, substituting (x, y) by (p, q) in both the equations, coordinates of P (p, q) can be evaluated by the formula shown in equation (1) in Figure 4.3.

A node, which is two hops away from primary reference nodes, cannot calculate its coordinates from the above parameters, as there is no direct link between the reference nodes and that node. So, in that case, a two-hops-away node should use nodes like P as its reference nodes to calculate its coordinates. Thus, P will act as a secondary reference node in this case. But before that, the secondary reference node like P has to synchronize its antenna with primary reference nodes so that it can share the common frame of reference and help other distant nodes to calculate their location with respect to the same frame of reference. So, after synchronization of antenna, P can volunteer itself as a secondary reference node.

4.3.2.1 Synchronization of an Antenna by a Nonprimary Reference Node

We will illustrate this problem and its possible solution considering the ESPAR antenna as our directional antenna model, which has 12 predefined beam patterns, identified by beam pattern 0 to 11 along a counterclockwise direction. Let us assume that a reference node R1

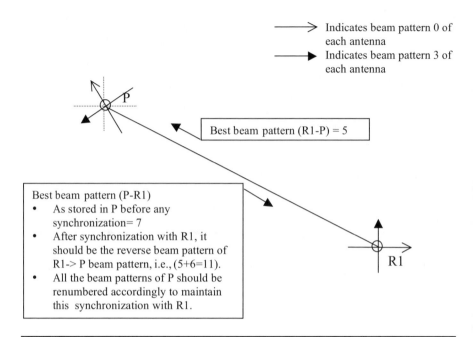

Figure 4.4 Illustration of the synchronization of antenna of P with reference node R1.

has aligned its 0th beam toward the east and the 0th beam of another node, say P, is set at an angle of 120° with respect to the 0th beam of R1 (Figure 4.4). Now, the beam pattern of R1 to access P is, say 5. So, ideally, in the common frame of reference, P should access R1 at its reverse beam pattern 11 (= 5 + 6).

But initial alignment of the 0th beam of P is at an angle of 120° in the counterclockwise direction with respect to that of R1. So according to P, the best beam pattern to access R1 is recorded as 7 in its AST instead of 11 (Figure 4.4). This conflict of beam pattern occurs because the initial alignments of antennas in P and R1 are different. So, P will extract the best beam pattern from R1 (or R2) from the beacon sent by R1 (or R2) according to their antenna alignments. But the original alignment of P may not be the same as the alignment of node R1 (or R2). So, P has to map that beam pattern according to the alignment of reference node R1 (or R2) in order to synchronize its frame of reference with primary reference nodes. But this does not mean that P has to ensure the similar physical alignment of antennas as in R1. So, P has to map its current beam patterns to synchronized beam patterns through some mechanism so that from then onward it can send beacons specifying the synchronized beam patterns instead of its original alignment of beam pattern. In other words, P renames its beam patterns without physically changing the orientation of its antenna.

This renaming is done through a mapping table. Thus, instead of physically aligning the antenna of a node, the mapping table logically aligns its original beam patterns to a common reference frame.

4.3.2.2 Formation of the Post-Synchronization Mapping (PM) Table

1. A node, say P, will first find out the beam pattern from which it is getting the strongest beacon signal from a neighboring reference node R1. This is the best beam pattern to access P by R1. This information is obtained from the NLST of P. Let this be x.
2. Then P will derive the reverse pattern of x. It can be done as follows:
   Because ESPAR has 12 beam patterns,
   Reverse beam pattern of $x = x + 6$, if $0 \leq x \leq 5$
   Reverse beam pattern of $x = x - 6$, if $6 \leq x \leq 11$
3. P will now consult its AST to find out its best-recorded beam pattern to access R1. Let this be y.
4. Then P will match the reverse beam pattern with y to form its post-synchronization mapping table as follows:
   a. P will first find out the difference obtained from the calculated value of the reverse beam pattern (from P to R1 as calculated using step 2) and the best beam pattern to access R1 as mentioned in its AST (y). This difference is called offset.
   b. For each beam pattern, P will then add this offset value to get the corresponding post-synchronized beam pattern as follows:
      i. If the resultant value is less than or equal to 11, then the resultant value will be stored in the PM table as the post-synchronized beam pattern of a corresponding pattern.
      ii. But if the resultant value is greater than 11, then (resultant value-11) it will give us the actual post-synchronized beam pattern of a corresponding pattern.

Let us consider the example shown in Figure 4.4.

- P has received the strongest signal from R1 at an angle, say 5.
- It will then calculate the reverse beam pattern of 5 as $5 + 6 = 11$.
- P will then find out the best beam pattern to access R1 from its AST as 7.

Table 4.1 Post-Synchronization Mapping Table at P

Naming of Beam Pattern before Synchronization	Post-Synchronized Naming of Beam Patterns
7	11
8	0
9	1
10	2
11	3
0	4
1	5
2	6
3	7
4	8
5	9
6	10

- Offset (difference between calculated reverse beam pattern P–R1 and beam pattern P–R1 as mentioned in AST) is $(11 - 7 = 4)$.
- Now it will form the post-synchronization mapping table in Table 4.1 according to the rules described above.

4.3.3 Estimating Location of a Node Multi-Hop Away from Reference Nodes

A node (say Q) that is multi-hop (say two hops) away from primary reference nodes R1 and R2 cannot calculate its coordinate directly using the process described in section 4.3.2 because there is no direct link between the reference nodes R1, R2, and the node Q. In that case, Q should use nodes like P and S as shown in Figure 4.5 as its reference nodes to calculate its coordinates. Thus, P and S will act as secondary reference nodes in this case. But before that, the secondary reference node like P has to synchronize its antenna with primary reference nodes using the scheme described in Section 4.3.2 so that it can share the common frame of reference and help other distant nodes like Q to calculate their locations with respect to the same frame

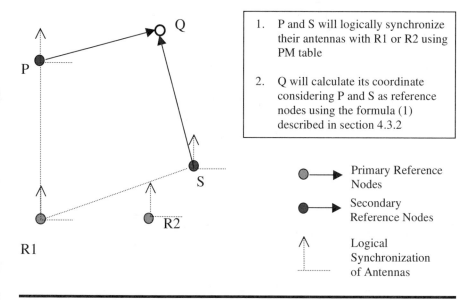

Figure 4.5 Location estimation by a node Q with respect to secondary reference nodes P and S.

of reference. So, after the formation of the post-synchronization mapping table at P (shown in Table 4.1), it can volunteer itself as a secondary reference node. Then Q can use the position of two neighboring secondary reference nodes (like P, S) and the corresponding angle of arrival (AoA) of signals from them to calculate its coordinate as per the formula shown in Section 4.3.2.

4.4 Implementation Results

We studied our proposal on a testbed in two steps. In the first step, we studied the effectiveness of location tracking with two nodes only in different environments (Figures 4.6(a), 4.6(b), 4.6(c)). Subsequently, as a second step, we set up a network of five nodes (Figure 4.7) and tested both location tracking and location estimation.

4.4.1 Two-Node Setting: Evaluating Location Tracking

Here node 1 is placed such that it is accessible by node 2 with beam pattern 3 of node 2. Then we started the location tracking in three environments: (1) in an anechoic chamber, (2) in a lounge during

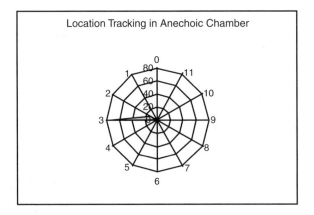

Figure 4.6(a)   Node 2 tracking node 1 with 80 percent accuracy.

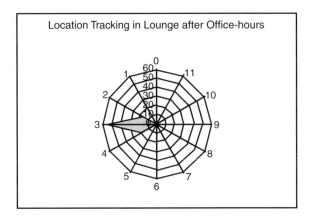

Figure 4.6(b)   Node 2 tracking node 1 with 50 percent accuracy.

office hours when there is movement of people, and (3) in the same lounge after office hours when there is no movement of people. Figures 4.6(a), 4.6(b), and 4.6(c) show the results. We took 100 samples in each case to track the location of one node by another and counted the number of occurrences of each beam pattern of node 2 as the best tracking direction of node 1. The number of occurrences of each beam pattern is plotted in Figures 4.6(a), 4.6(b), and 4.6(c). Most of the time the tracked location is beam 3, as expected. But at times, the tracked location is adjacent beams (i.e., 2 or 4) or other beams due

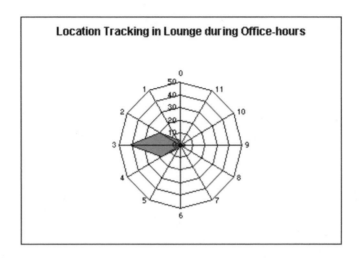

Figure 4.6(c) Node 2 tracking node 1 with 40 percent accuracy.

to multipath effects. Because beam 3 is the correct beam pattern of node 2 to access node 1, in an anechoic chamber the tracking accuracy is 80 percent. Tracking accuracy in the lounge after office hours is 50 percent and during office hours is 40 percent. So, in normal environments, location tracking can be done accurately only if we take a number of samples.

4.4.2 Five-Node Setting: Single-Hop Location Estimations with Two Reference Nodes

As a next step, we set up a network of five nodes in the lounge as shown in Figure 4.7. Tracking of N3, N4, and N5 by reference nodes N1 and N2 is shown in Figure 4.8.

As shown in Figure 4.7, 0th beam pattern of both the reference nodes N1 and N2 are oriented toward X.

Based on these, nodes N3, N4, and N5 compute their coordinates. This is done using the following formula (where p is the x coordinate and q is the y coordinate of the target node):

$$p = ((y2 - y1) + x1 \tan \alpha - x2 \tan \beta) / (\tan \alpha - \tan \beta)$$

$$q = ((x1 - x2) \tan \alpha \tan \beta + (y2 \tan \alpha - y1 \tan \beta)) / (\tan \alpha - \tan \beta)$$

The actual coordinates of the nodes are shown in small circles and are as follows (in meters):

N1 (Reference node 1) = (0, 0)
N2 (Reference node 2) = (0, 3.5)
N3 = (0, 5)
N4 = (3.5, 6.5)
N5 = (5.5, 2.5)

The tracking of N3, N4, and N5 by reference nodes N1 and N2 is shown in Figure 4.8 and the corresponding tracking angles are

$\angle N_3N_1N_2 = 90°$
$\angle N_4N_1N_2 = 60°$
$\angle N_5N_1N_2 = 30°$
$\angle N_3N_2X = 120°$
$\angle N_4N_2X = 90°$
$\angle N_5N_2X = 60°$

$\longrightarrow X$ (0th beam direction of N1 and N2)

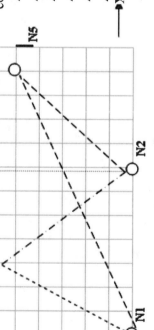

Figure 4.7 Single-hop location estimations of N3, N4, and N5 with two reference nodes N1 and N2. The computed coordinates are shown in black circles.

In the case of N3, $\alpha = 90°$ and $\beta = 120°$,
y1 = y2 = x1 = 0, x2 = 3.5; so, N3 (p, q) = N3 (0, 6)

In the case of N4, $\alpha = 60°$ and $\beta = 90°$,
y1 = y2 = x1 = 0, x2 = 3.5; so, N4 (p, q) = N4 (3.5, 6)

In the case of N5, $\alpha = 30°$ and $\beta = 60°$,
y1 = y2 = x1 = 0, x2 = 3.5; so, N5 (p, q) = N5 (5.2, 3)

The computed coordinates are shown in black circles in Figure 4.7.

4.4.3 Five-Node Setting: Multi-Hop Location Estimations with Secondary Reference

To have a situation where multiple nodes could be placed closely together and the signals from neighboring nodes filtered appropriately to give a different topology, we used packet-filtering software similar to Macfilter[12] or Powerwave.[13] This will prevent packets from being transmitted directly from one node to the other, depending on our desire. Using this mechanism, we isolated node N4 from N1 and N2 to have a multi-hop configuration. In other words, N4 is no longer a neighbor of two reference nodes N1 and N2, and N4 estimates its coordinate treating N3 and N5 as secondary reference nodes. Original alignment of 0th beam pattern of N3 was at an angle 180° with respect to N1. So, the calculated offset value in this case will be 6. The best beam pattern to access N4 as recorded in the AST of N3 is 7. So, after synchronization using the rule described in the previous section, beam 7 of N3 should be renumbered as 1 (= 7 + 6). But no such synchronization is required at N5 as the original alignments of N5 and N1 are the same. Here, N3 and N5 first align their 0th beam pattern toward the X axis of reference nodes N1 and N2. This is a process of synchronization of beam patterns of secondary reference nodes, as illustrated earlier.

The computed position of N4 using N3 and N5 as secondary reference points is shown as a small black square in Figure 4.9.

X coordinate of N4 = ((y2 − y1)
+ x1 tan α − x2 tan β) / (tan α − tan β) = 2.6

Y coordinate of N4= ((x1 − x2) tan α tan β
+ (y2 tan α − y1 tan β)) / (tan α − tan β) = 7.5

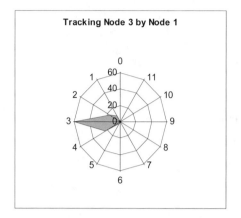
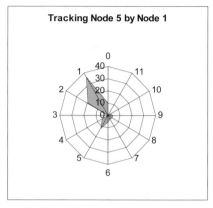
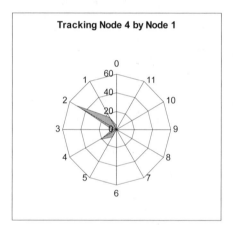
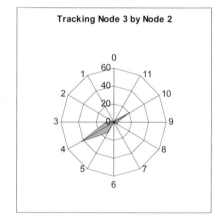
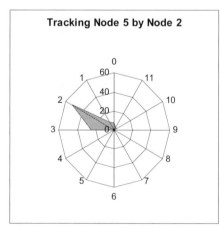
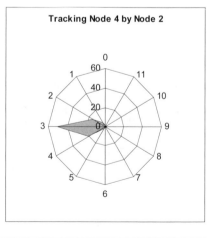

Figure 4.8 Tracking beam directions of N1 and N2 while tracking N3, N4, and N5.

Subsequently, the tracking angle of N4 with respect to N3 (α) is found to be 30° and that of N4 with respect to N5 (β) is found to be 120°. Computed coordinate of N3 is (0, 6) and that of N5 is (5.2, 3) with reference to primary reference nodes N1 and N2. Based on this, N4 calculates its coordinates taking N3 and N5 as its reference points as follows:

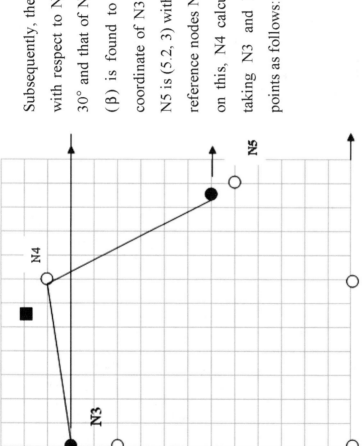

Figure 4.9 Multi-hop location estimations of N4 with two secondary reference nodes N3 and N5. The computed position of N4 using N3 and N5 as secondary reference points is shown as a small black square.

4.5 Error in Location Estimation: A Simulation Study

There will be two types of errors that will occur in location estimation: first, an estimation error due to the error in location tracking, and, second, an estimation error due to beamwidth, which limits the estimation of AoA to discrete values as 30°, 60°, and so on.

The experimental results of location tracking shown in Figures 4.6 and 4.8 depict that the accuracy of location tracking depends on several external factors like environment, multipath effects, and so on. Empirical results indicate that it is possible to get the correct beam pattern to access a node if a considerable number of beacons are observed. In a line-of-sight communication environment, the most frequently occurred beam pattern would be the correct beam pattern to access a node even in the presence of multipath effects.

To minimize the error due to tracking we observed 100 samples and calculated the frequency of occurrence of beam patterns. Therefore, the beam pattern with maximum frequency is selected as the best beam pattern of a node to access a neighbor. The sample size has been determined using the following formula:

n = $pq/(SE)^2$ where,
n = sample size
p = probability of getting accurate beam pattern to access a node
q = $1 - p$
SE = Standard error of the proportion

With confidence interval 10 percent and the confidence level of 95 percent, the standard error is:

$$SE = .10/1.96 = .05$$

We assumed p as .5 and q as .5. So, by replacing $p = .5$ and $q = .5$ and $SE = .05$ we get $n = 100$.

The estimation error due to beamwidth is shown in Figure 4.10 where two reference nodes R1 and R2 are trying to track the location of N. Even if the AoA estimation is correct, the shaded region is the region of error and N may reside anywhere in this region. This region of error is proportional to beamwidth and the product of d1 and d2. Reducing beamwidth will definitely improve the estimation error. At the same time, if d1 or d2 increases, the estimation error increases. When we use a multi-hop estimation technique, the error is compounded,

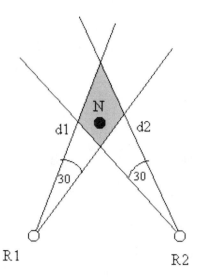

Figure 4.10 Error estimation.

because the computed coordinates of secondary reference nodes themselves have some error. However, if a node is a single hop away from primary reference nodes with large d1 and d2, it may be better to compute the coordinate of that node in multi-hop, because that would reduce the value of d1 and d2. Moreover, in n-hop scenario, as the location estimation at each hop can produce error with the same probability in every direction, the distribution of the direction of estimation error will be uniform.[11] As a result, the error propagation in case of multi-hop location estimation may not be very significant.

4.6 Discussion

In this chapter, we proposed a mechanism for location tracking and location estimation of nodes in ad hoc networks using a directional antenna and evaluated the effectiveness of the scheme on a testbed. Our focus is not so much on accuracy of location estimation. Each node in the system has to do neighborhood location tracking to implement directional MAC and routing protocols. We have just used this knowledge for approximate location estimation without using additional hardware. The possibility of using multiple reference nodes to track the location more accurately even in non-line-of-sight environments will be a scope for future work.

References

1. Ueda, T. et al., A Rotational Sector-based, Receiver-oriented mechanism for Location Tracking and Medium Access Control in Ad Hoc Networks Using Directional Antenna, in *Proc. IFIP Conference on Personal Wireless Communications*, Venice, Italy, September 23–25, 2003.
2. Streenstrup, M.E., Neighbor Discovery among Mobile Nodes Equipped with Smart Antennas, www.wireless.kth.se/adhoc03/Proceedings/mssession6-1.pdf.
3. Nasipuri, A. and Li, K., A Directionality-based Location Discovery Scheme for Wireless Sensor Networks, in *Proc. 1st ACM International Workshop on Wireless Sensor Networks and Applications,* Atlanta, Georgia, September 2002.
4. Lee, J. et al., Location Sensing Protocol Utilizing Directional Antenna Systems in Millimeter-wave Broadband Ad Hoc Networks, technical report, SNU-CSE-MMLAB-LSDA version 1, September 2003.
5. Bulusu, N., Heidemann, J., and Estrin, D., GPS-less Low Cost Outdoor Localization for Very Small Devices, in *IEEE Personal Communications Magazine,* October 2000, p. 28.
6. Harter, A. et al., The Anatomy of a Context-aware Application, in *Proc. CM/IEEE MOBICOM 1999,* p. 59.
7. Ward, A., Jones, A., and Hopper A., A New Location Technique for the Active Office, in *Proc. IEEE Personal Communications*, Vol. 4, No. 5, October 1997, p. 42.
8. Priyantha, N., Chakraborthy, A., and Balakrishnan, H., The Cricket Location Support System, in *Proc. ACM/IEEE International Conference on Mobile Computing and Networking (MOBICOM),* August 2000.
9. Hightower, J., Want, R., and Borriello, G, SpotON: An Indoor 3D Location Sensing Technology Based on RF Signal Strength, UW CSE 2000-02-02, University of Washington, Seattle, February 2000.
10. Griswold, W.G. et al., ActiveCampus: Sustaining Educational Communities through Mobile Technology, University of California, San Diego, Department of Computer Science and Engineering technical report, 2002.
11. Capkun, S., Hamdi, M., and Hubaux, J., GPS-Free Positioning in Mobile Ad Hoc Networks, in *Proc. 34th Annual Hawaii International Conference on System Sciences (HICSS-34),* Vol. 9, Maui, Hawaii, January 3–6, 2001.
12. Maltz, D.A., Broch, J., and Johnson, D.B., Quantitative Lessons from a Full-scale Multi-hop Wireless Ad Hoc Network Testbed, in *Proc. IEEE Wireless Communications and Networking Conference*, September 2000.
13. Chin, K.-W. et al., Implementation Experience with MANET Routing Protocols, in *Proc. ACM SIGCOMM Computer Communication Review,* Vol. 32, No. 5, November 2002, p. 49.

Chapter 5
A Routing Strategy for Effective Load Balancing Using Smart Antennas

5.1 Introduction

Several MAC protocols with directional antennas have been proposed in the context of ad hoc networks in order to improve media utilization with increased number of simultaneous communications. However, even if we have an efficient directional MAC protocol, it alone would not be able to guarantee good system performance, unless we have a proper routing strategy in place that exploits the advantages of a directional antenna. In this chapter, we propose a notion of zone-disjoint routes and define maximally zone disjointness as route selection criteria. Subsequently, an adaptive routing strategy is proposed that exploits the advantages of directional antennas in ad hoc networks through the selection of maximally zone disjoint routes. Zone-disjoint routes would minimize the effect of route coupling in wireless media by selecting routes in such a manner that data communication over one route will minimally interfere with data communication over the others. The proposed routing strategy ensures effective load balancing in wireless media and is applied to design and implement both single-path and multipath routing protocols on ad hoc networks with directional antennas.

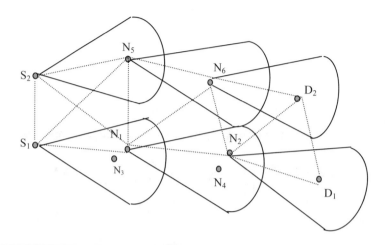

Figure 5.1 Zone-disjoint communications between S_1–D_1 and S_2–D_2 with directional antennas.

Let us consider the scenario in Figure 5.1, where source S_1 is communicating with destination D_1 through N_1 and N_2. At the same time, suppose another source S_2 also wants to communicate with destination D_2. Suppose there are three possible paths: {S_2, N_1, N_2, D_2}, {S_2, N_3, N_4, D_2}, and {S_2, N_5, N_6, D_2}. If S_2 uses the first path that overlaps with the path used by S_1, then simply using a directional antenna cannot improve the routing performance. If S_2 uses the second path, then also routing performance will deteriorate because of the phenomenon known as route coupling.[1-3] Route coupling occurs when two routes are located physically close enough to interfere with each other during data communication. As a result, the nodes in those two routes are constantly contending for access to the medium they share. In Figure 5.1, because the nodes belonging to these two routes are within the transmission zone of one another (even if we use directional antennas, as shown), these two communications cannot happen simultaneously: N_1 and N_3 cannot receive data simultaneously from S_1 and S_2, respectively; similarly, N_2 and N_4 cannot receive data simultaneously from N_1 and N_3, respectively.

The routing performance between any source and destination does not depend only on the congestion characteristics of the nodes in that path. Pattern of communication in the neighborhood region will also contribute to this delay. This is a phenomenon known as route coupling. Thus, even if {S_1, N_1, N_2, D_1} and {S_2, N_3, N_4, D_2} are node-disjoint, routing performance will deteriorate in this context, even if we use directional antennas.

The impact of directional antennas on routing would be visible if S_2 selects the third path, or $\{S_2, N_5, N_6, D_2\}$. The two routes $\{S_1, N_1, N_2, D_1\}$ and $\{S_2, N_5, N_6, D_2\}$ are coupled with each other, if we use omni-directional antennas (as shown with dotted lines in Figure 5.1). But they are completely decoupled, if we use directional antennas, as shown in Figure 5.1. These two routes are said to be zone-disjoint, because data communication over one path will not interfere with data communication over the other path.

It is evident from the above discussion that zone-disjoint routes ensure effective load balancing in the network. We have developed a routing strategy for effective load balancing through the selection of zone-disjoint routes, exploiting the capacity of directional antennas. In recent times, some researchers have developed routing strategies with load balancing in ad hoc networks using omni-directional antennas.[4,5] They consider intermediate node routing loads or nodal activity information of all nodes as the primary route selection metric. The application of multipath routing techniques in mobile ad hoc networks has also been explored to reduce end-to-end delay and perform load balancing.[2,6] But none of the proposals have explored the route-coupling phenomenon for effective load balancing.

Distributing the routing tasks evenly throughout the network has two major advantages in this context. First, it prevents loads from concentrating on a set of nodes and spreads them among other nodes in a uniform manner, thereby reducing the possibility of power depletion of a set of heavily used nodes. Second, it distributes traffic all over, thus reducing congestion and improving network performance. Most of the current proposals on load balancing in this context would help to distribute traffic all over the network and can achieve the first advantage as mentioned above. However, because of route coupling in wireless media, as shown in Figure 5.1, distribution of traffic alone cannot guarantee improved throughput. In Figure 5.1, $\{S_1, N_1, N_2, D_1\}$ and $\{S_2, N_3, N_4, D_2\}$ are node-disjoint and consequently satisfy the criteria for load balancing. But, because they are coupled with each other, end-to-end delay will increase. The larger the degree of coupling, the larger will be the average end-to-end delay for both paths.[1] This is because two paths have more chances to interfere with each other's transmission due to the broadcast feature of radio propagation. That is why it is important to discover zone-disjoint routes for effective load balancing.

But getting zone-disjoint or even partially zone-disjoint paths using omni-directional antennas is difficult because the transmission zone of an omni-directional antenna covers all directions. A directional antenna has a reduced transmission beamwidth compared to an omni-directional antenna. In our example, two routes $\{S_1, N_1, N_2, D_1\}$ and $\{S_2, N_5, N_6, D_2\}$ are zone-disjoint, only if we use a directional antenna. It has been shown that it is much easier to get zone-disjoint routes and, consequently, the effect of route coupling can be drastically reduced, if we use a directional antenna instead of an omni-directional antenna with each user terminal forming an ad hoc network.[3,7]

However, zone-disjointness alone is also not sufficient for performance improvement. Path length is also another important factor. A longer path with a larger number of hops (H) will increase the end-to-end delay and waste network bandwidth, even in the context of zone-disjointness. So, it is imperative to select maximally zone-disjoint shortest paths.

The routing scheme discussed so far employs single-path routing. However, instead of using a single path for each communication, it is possible to improve network throughput by splitting the total volume of data into separate blocks and sending them via selected multiple paths from s to d, which would eventually reduce congestion and end-to-end delay.[8] But it has also been shown that deployment of multiple paths may not necessarily result in a lower end-to-end delay.[2] It was argued that the network topology and channel characteristics (e.g., route coupling) can severely limit the gain offered by multipath routing strategies. In this chapter, we use the same notion of zone-disjoint shortest path routing scheme for multipath routing and show the performance improvement of multipath routing over single-path routing under low traffic loads.

In our implementation, each node is aware of its neighborhood and communications going on in its neighborhood at that instant of time. Moreover, it keeps track of directional access information for its neighborhood nodes. This helps each node to determine the best possible direction of communication with any of its neighbors. This status information from each node is propagated periodically to its neighbors; each of them assimilates this information and further propagates to its neighbors at a periodic interval. Thus, information percolated throughout the network would help each node to capture the approximate network status periodically without generating a lot of control traffic. Thus, each node becomes topology-aware and aware of communications going on in the network, although in an approximate manner. We have proposed a table-driven routing protocol for both single-path and multipath routing. We have defined and developed a metric for measuring maximally zone-disjointness and used it as route selection criteria for load balancing. However, because network awareness at each node is only a perception about network status rather than *actual* network status, each intermediate node adaptively corrects and modifies routing decisions during routing. We have evaluated the effectiveness of single-path and multipath routing schemes on the QualNet network simulator with AODV[9,10] (as in QualNet) as a benchmark. Our proposed mechanism shows four to five times performance improvement over AODV, thus demonstrating the effectiveness of this proposal.

5.2 System Description

5.2.1 Some Important Definitions

Definition 1

When a node n forms a transmission beam at an angle α and a beamwidth β with a transmission range R, the coverage area of n at an angle α is defined as transmission_zone$_n$ (α,β,R) (Figure 5.2) of node n. It implies that if a node m is within the transmission_zone$_n$ (α,β,R) and m is in receive mode, then whenever n transmits a message at that transmission angle α with respect to n and beamwidth β and transmission range R, it will be received by m. When node m moves out of the transmission_zone$_n$ (α,β,R), the connectivity between n and m is lost. Because transmission beamwidth β and transmission range R are fixed here in our study, we will refer to transmission_zone$_n$ (α,β,R) as transmission_zone$_n$ (α) in subsequent discussions.

Definition 2

We define neighbors of n (G^n) as a set of nodes within the omnidirectional transmission zone of n.

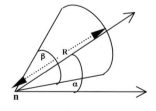

Figure 5.2 Transmission zone$_n$ (α,β,R).

Definition 3

A subset of G^n, $G^n_\alpha \in G^n$, is defined as the directional neighbors of n, where the nodes in G^n_α lie within its transmission_zone$_n$ (α).

Definition 4

Communication-ID c is essentially a unique ID that specifies a source-destination pair for which the communication is on. In the case of multipath communication from a source to a destination, a sub-ID of that communication-ID represents each of the multipath flows.

Definition 5

The active node list [ANL(t)] is a set of nodes in the network actively participating in any communication process at an instant of time t. Each active node in the list is associated with a set C of communication-IDs for which it is active.

Definition 6

Active directional neighbors of node n at transmission_zone$_n$ (α) [ActG$^n_\alpha$ (t)] is a set of nodes within the transmission_zone$_n$ (α) that are actively participating in any communication process at that instant of time (i.e. belongs to ANL(t)). So, ActG$^n_\alpha$ (t) = G^n_α (t) $\cap$ ANL(t).

Definition 7

Route coupling is a phenomenon of wireless media that occurs when two routes are located physically close enough to interfere with each other during data communication.[1,2]

In Figure 5.3, let n_1–n_7 and n_2–n_6 be the two communications (represented by communication-IDs c_1 and c_2, respectively) present in a network at any instant of time. It is evident from the figure that the transmission_zone$_{n1}$ (α_1) used by c_1 is interfering with transmission_zone$_{n2}$ (α_2) used by c_2, which restricts the possibility of simultaneous communications $n_1 \rightarrow n_3$ and $n_2 \rightarrow n_4$. The contention between two communications for accessing the medium will result in high end-to-end delay and the performance of each interfering communication will suffer. So, it is evident that elimination of route coupling between two interfering routes will definitely lead to improved network performance.

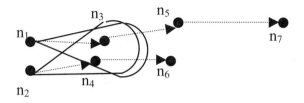

Figure 5.3 Route coupling causes contention in wireless medium.

In order to select the best path for a communication, one has to measure the degree of coupling of each available path for that communication with the paths used by other ongoing communications (active paths) in the network and select the path with the lowest degree of coupling. Correlation factor η is used to measure route coupling.[1,2] It has been shown that the larger the correlation factor, the larger will be the average end-to-end delay for both paths.[1] This is because two paths with larger correlation factors have more chances to interfere with each other's transmissions due to the broadcast feature of radio propagation.

Definition 8

The correlation factor of node n_i in a path P for communication-ID c [$\eta^{ni}_c(P)$], where n_j is the next hop from n_i in path P and $\alpha(n_i \to n_j)$ is the transmission zone formed by n_i toward n_j in order to communicate with n_j, is defined as the sum of the number of communication-IDs handled by each active directional neighbor of node n_i at transmission_zone$_{ni}$ ($\alpha(n_i \to n_j)$) excluding the current communication-ID c. So, $\eta^{ni}_c(P) = \Sigma_{\forall n \in \text{ActG-ni-}\alpha(ni \to nj)(t)} (|C - c|)$. Informally speaking, the correlation factor of a node along a specified path measures the degree of coupling of that node and its directional neighbors with other active communications in that given communication direction. For example, if the correlation factor of node n_i in path P along transmission zone $\alpha(n_i \to n_j)$ is zero, it implies that node n_i can transmit to n_j at transmission zone $\alpha(n_i \to n_j)$ without affecting any other communication. On the other hand, if n_i has two active directional neighbors at transmission_zone $\alpha(n_i \to n_j)$, with one handling two communications and the other handling four communications and if one of them is handling communication-ID c, then [$\eta^{ni}_c(P)$] will be 2 + 4 − 1 = 5. This implies that node n_i can transmit to n_j at transmission_zone $\alpha(n_i \to n_j)$ but will affect other communications with a degree of coupling 5. It is to be noted that, if an active directional neighbor of a node n_i is active with its own communication-ID c, then n_i ignores the activity status of that node for that communication-ID c while calculating $\eta^{ni}_c (P)$.

Definition 9

Correlation factor η of path P for communication-ID c [$\eta_c (P)$] is defined as the sum of the correlation factors of all the nodes in path P. So, $\eta_c (P) = \sum_{\forall ni \in P} (\eta^{ni}_c (P))$. When $\eta_c (P) = 0$, path P is said to be zone-disjoint with all other active paths, where active paths are those paths participating in the communication process at that instant of time. Otherwise, the path P is η-related with other active paths. Correlation factor is used to measure route coupling. As an example, let us refer back to Figure 5.1 and let us assume that each node is handling one communication only. Initially, source S_1 is communicating with destination D_1 through N_1 and N_2. So, ANL(t) contains {S_1, N_1, N_2, D_1}. Now, S_2 wants to communicate with D_2 and selects a path P = {S_2, N_5, N_6, D_2}. Let us first consider the case with omni-directional antennas (Figure 5.1). Both S_1 and N_1 are within the transmission zone (which is 360 degrees in this case) of S_2. So, $\eta^{S2} (P) = 2$. Because S_1 and N_1 are within the omni-directional transmission zone of N_5, $\eta^{N5} (P) = 2$. Similarly, $\eta^{N6} (P) = 2$. So, $\eta (P) = 6$, when we use omni-directional antennas. When we use directional antennas, the transmission zones formed by S_2, N_5, and N_6 do not contain any node from ANL(t), as shown in Figure 5.1. So, $\eta (P) = 0$, when we use directional antennas.

Definition 10

For a path P, $\eta (P) = 0$ implies that any communication along path P will not disturb any ongoing communication process at that instant of time. Such a path is known as a zone-disjoint path with respect to all other active paths used by other ongoing communications at that point of time. Otherwise, the path P is said to be η-related with other active paths. We propose to use two metrics in our route selection criteria: correlation factor and propagated hop count. As explained earlier, correlation factor of a route is inversely related to zone disjointness of that route with respect to other active routes. At the same time, hop count of the selected route is also another concern in this context. Otherwise, under some communication scenario, it may so happen that, for a particular destination, each intermediate node tries to select a route avoiding the existing communication zones and ultimately ends up traversing the entire network in search of a zone-disjoint route. So, minimization of both the correlation factor and propagated hop count will give rise to a maximally zone-disjoint shortest route.

5.2.2 Network Awareness

In our proposal, each node keeps certain neighborhood status information dynamically so that each node is aware of its neighborhood and communications going on in its neighborhood at that instant of time. This status information from each node is propagated periodically throughout the network. This would help each node to capture the approximate network status periodically that in turn helps them to become topology-aware and aware of communications going on in the network, although in an approximate manner. With this network-status information, each intermediate node adaptively computes maximally zone-disjoint routes (single path or multipath) toward its destination, as will be discussed later.

Each node n in the network has the following four pieces of network-status information:

- Angle signal table (AST_n): In order to track the direction of its neighbors, each node n periodically collects its neighborhood information and forms an angle signal table (AST). For each neighbor $m \in G^n$, $AST_n(t)$ of node n specifies the maximum strength of radio connection, $SIGNAL^\theta_{n,m}(t)$, as perceived by n, at a particular direction. Thus, $SIGNAL^\theta_{n,m}(t)$ is the maximum strength of received signal at node n from its neighboring node m at an angle θ with respect to n and as perceived by n at any point of time t. The AST of node n will help us to determine the best possible direction of communication with any of its neighbors.
- Neighborhood active node list ($NANL_n$): NANL at node n contains the communication-activity-status of its neighbors. In other words, if any of the neighbors of node n is actively participating in a communication process or is inactive, node n records that information in a list called the neighborhood active node list ($NANL_n[t]$). This helps a node to become neighborhood-communication-aware.
- Active node list (ANL_n): It contains the perception of node n about current communication activities in the entire network.

It is a list in node n containing all active nodes in the network, as perceived by n at that instant of time.
- Global link-state table ($GLST_n$): It contains the network topology information as perceived by n at that instant of time.

Each node broadcasts its ANL at a periodic interval, say T_A. Broadcast of ANL serves two purposes: when a node n receives ANL from all its neighbors (say nodes i, j, and k):

- Node n forms the AST_n to include nodes i, j, and k as its neighbors and records the best possible direction to communicate with each of them.
- Node n also records the communication activity status of node i, and similarly for other neighbors, thus forming its own NANL, and subsequently upgrades its ANL.

Each node broadcasts its GLST at a periodic interval, say T_G. When a node n receives GLST from its neighbors, it updates its own GLST, as will be illustrated later. In our system, $T_A = 500$ msec and $T_G = 10$ seconds. The reason for broadcasting two packets at two intervals is as follows: ANL captures the communication activity and once a communication starts, immediately a set of nodes will be affected. So, ANL needs to be propagated faster than GLST. Moreover, ANL serves as a beacon. So, by the faster propagation of ANL, not only the critical information of active nodes can be percolated faster, but also accurate neighborhood information (direction, signal level) can be obtained. Because we are implementing the fisheye concept,[11] accurate neighborhood information is required faster. On the other hand, GLST is the global information about connectivity of all nodes. It reflects the change of topology with respect to physical mobility (which is much slower compared to signal propagation). Moreover, GLST at any node need not be so accurate. That is why GLST, the larger packet, propagates slowly and ANL, the smaller packet, propagates faster.

5.2.2.1 Formation of AST and NANL

A node, say n, forms its AST_n incrementally on receiving ANL packets from any of its neighbors, say m. Because n will receive an ANL packet from m by setting its antenna at a particular angle, it knows the best possible direction to communicate with m and the maximum strength of radio connection, $SIGNAL^\theta_{n,m}(t)$, as perceived by n, at that particular direction Here, we assume symmetric links.

Any node, say n, forms its $NANL_n$ with its own activity status first. Whenever n needs to issue an RTS, indicating that it has a desire to communicate, it sets itself as active node; whenever it is not issuing RTS for a threshold period of time, it sets itself as inactive. Additionally, whenever node n receives an RTS from its neighbor, say m, it sets m as active node in its $NANL_n$. Whenever node m deactivates itself, this information reaches n through periodic broadcast of an ANL from m. On receiving that, n sets m as inactive in its $NANL_n$.

5.2.2.2 Formation of ANL

Each node periodically broadcasts its active node list, which contains its perception about communication activities in the network. On receiving periodic ANL from different nodes, each node combines them to form a revised ANL and waits for a periodic interval to broadcast it to its neighbors.

At each node, ANL first gets updated by the neighborhood active node list (NANL) of that node. Initially, when the network commences, all the nodes are just aware of the activity status of their own neighbors and are in a don't-know state regarding the other nodes in the system. Periodically, each node broadcasts its ANL to its neighbors. With this periodic update from its neighbors about their neighbors, the nodes slowly get activity information about the other nodes in the network. Thus, each node updates its own ANL based on received update messages from other nodes.

A major aspect underlying the infiltration of network status information into mobile nodes is that the information carried must be recognized with a degree of correctness. Because the propagation of updates from different nodes is asynchronous, it becomes imperative to introduce a concept of recency of information.[12,13] For example, let us assume two ANL packets A_1 and A_2 arrive at node n, both of them carrying information about node m, which is a multi-hop away from node n. In order to update the information at node n about node m, there has to be a mechanism to find out who carries the most recent information about node m: A_1 or A_2?

To implement this, we have used the concept of recency and a mechanism to increment it appropriately. If two update messages have a set of data concerning the same node, say node n, then the update message carrying the higher recency value of node n has more current information about it. The structure of a typical ANL at a node n is given in Table 5.1.

Table 5.1 The Structure of an ANL

Nodes	n_1	n_2	...	n_N
Recency	R_1	R_2	...	R_N
State	S_1	S_2	...	S_N
Communication-ID	C_{11}	C_{21}	...	C_{N1}
	C_{12}	C_{22}	...	C_{N2}
	C_{13}	C_{23}	...	C_{N3}

Here, R_i is the recency of node n_i in a network of N nodes and S_i denotes the corresponding activity status of each node, which can be either 0 (inactive) or any positive value (active). Any value of S_i greater than 0 indicates the node to be active for S_i number of communications. C_{ij} denotes the communication-ID of j-th communication for which i-th node is active. Only the first three communication-IDs are propagated for each active node. If node i is inactive, $C_{ij} \: \forall \: j, j \in \{1, 2, 3\}$ is null.

5.2.2.3 Formation of GLST

Each node maintains a global link-state table (GLST) to capture network connectivity information. At each node, GLST first gets updated by the AST of that node. Initially, when the network commences, all the nodes are just aware of their own neighbors and are in a don't-know state regarding the other nodes in the system. Periodically, each node broadcasts its GLST to its neighbors. With this periodic update from its neighbors about their neighbors, the nodes slowly get information about the other nodes and their neighbors. Thus, each node updates its own GLST based on received update messages from other nodes. By controlling the periodicity of updates, it is possible to control the update traffic in the network and the accuracy of network status information stored in each of the nodes. For example, if the propagation of update messages is too frequent, the control traffic will increase but the accuracy of network status information stored in each node will also be better. However, the network would never get flooded with propagation of updates. The maximum number of update packets in the network at any point of time is always less than the number of nodes in the network. In this case also, we need to implement the concept of recency as explained in the context of ANL propagation. This implies that if two GLST update messages have a set of data concerning the same node, say node n, then the update message

Table 5.2 The Structure of a GLST

Nodes	Recency	Neighbors
n_1	R_1	→ {..............}
n_2	R_2	→ {..............}
...	...	→ {..............}
n_i	R_i	→ {<$n_j, \alpha(n_i, n_j)$> <$n_k, \alpha(n_i, n_k)$>..........}
...	...	→ {..............}
n_N	R_N	→ {..............}

carrying the higher recency token value of node n has more current information about it.

As and when a node n receives GLST from other nodes, it updates its GLST. In order to do that, the recency tokens of all the nodes stored in the GLST of n and the recency tokens of all the nodes stored in the recently arrived update packet are compared. If the recency token of any node, say X, in GLST of n happens to be less than that in the update packet, then it is obvious that the update packet is carrying more recent information about node X. So, the entire information about node X in the GLST of node n is overwritten by the received information of X in the update packet. This step is performed asynchronously for all the update packets as they arrive at that host node n. This step helps node n to acquire all the recent information that it can gather from the update packets.

The mechanism does not guarantee that each node would know the exact status of the network. It is merely an awareness that helps each node to figure out the approximate status of the network. This is similar to the fisheye approach,[11] which helps to maintain accurate status information about the immediate neighborhood of a node, with progressively less accurate details as the distance increases. The structure of GLST at any node n is given in Table 5.2.

Here, R_i is the recency of node n_i in a network of N nodes and <$n_j, \alpha(n_i, n_j)$> denotes that n_j is a neighbor of n_i where $\alpha(n_i, n_j)$ indicates the transmission beam angle α at which n_i can best communicate with n_j.

5.3 Network-Aware Routing with Maximally Zone-Disjoint Shortest Path

As mentioned earlier, even if we have an efficient directional MAC protocol, it alone would not be able to guarantee good system performance unless we have a proper routing strategy in place that exploits the advantages of a directional antenna. Little work has been done on effective routing protocols using directional antennas. Two proposals have discussed on-demand routing using directional antennas.[14,15] Nasipuri et al. have proposed that a directional antenna be used to reduce search space during flooding of route-request packets to search a route in on-demand basis. However, effective use of a directional antenna during routing has not been discussed. Roy Choudhury and Vaidya proposed an on-demand directional-DSR scheme and a preliminary analysis has been done to identify the appropriateness of directional antennas in on-demand routing. In this chapter, a table-driven or proactive adaptive routing strategy is proposed that truly exploits the advantages of directional antennas in ad hoc networks through the selection of maximally zone-disjoint shortest routes. Zone-disjoint routes would minimize the effect of route coupling in wireless media by selecting routes in such a manner that data communication over one route will minimally interfere with data communication over the others. The proposed routing strategy ensures effective load balancing in wireless media and is applied to design and implement both single-path and multipath routing protocols on ad hoc networks with directional antennas.

5.3.1 Maximally Zone-Disjoint Shortest-Path Routing

Traditional routing schemes generally use a shortest-path routing protocol to improve the network performance in terms of throughput. But the selection of the shortest path for each communication is not at all sufficient to improve network performance under high-traffic scenarios. Several communications may use some common nodes while trying to route their data packets through shortest paths. Therefore, congestion may occur at those common nodes, which in turn decreases the network throughput. In the context of wireless environments, not only common nodes but also use of common zones during routing increases the end-to-end delay because of route coupling. So, instead of the shortest path, if a diverse zone-disjoint path (which is zone-disjoint with respect to the existing flow of traffic) could be selected for a new communication, avoiding the zones already involved in some other communications, then that would definitely improve the throughput by reducing the congestion.

At the same time, hop count of the selected diverse route is another concern in this context. Otherwise, under some communication scenario, it may so happen that, for a particular destination, each intermediate node tries to select a route avoiding the active zones and ultimately ends up traversing the entire network in search of a zone-disjoint route. To alleviate that problem we propose to use two metrics as route selection criteria: correlation factor and propagated hop count, as will be explained below. The correlation factor of a route is inversely related to zone disjointness of that route with respect to other active routes. So, minimizing correlation factor and minimizing propagated hop count will give rise to maximally zone-disjoint shortest paths.

Each node in the network uses its current network status information (approximate topology information and ongoing communication information) to calculate the *suitable next hop* for reaching a specified destination such that:

- The interference with the nodes that are already involved in some communication gets minimized.
- If a packet at an intermediate node has already traversed multiple hops, then shorter hop routes toward the destination get more preference.

In other words, initially when a packet is transmitted from the source it gives preference to the zone-disjoint path-selection criteria. If a packet already traversed multiple hops, then the progressively shortest hop route toward the destination will be selected. This adaptive route calculation mechanism guarantees the convergence of the proposed routing algorithm. We have used the following function to calculate the link weight that will ensure the selection of a lower η path for a low propagated hop count and the selection of a lower hop path for a higher propagated hop count.

Link-cost (n_i, n_j) during the current communication having communication-ID $c = \alpha + \eta_c^{ni} + \gamma H$ where,

α = Initial link weight (.01 in our case; $\alpha << \eta_c$ and $\alpha << H$, as will be explained in the following section)

η_c^{ni} = The sum of the total number of communications (excepting the current communication c) handled by each directional active neighbor in the directional zone $(n_i \rightarrow n_j)$, i.e., $\eta_c^{ni} = \Sigma_{\forall n \in ActG\text{-}ni\text{-}\alpha(ni \rightarrow nj)\ (t)} (|C - c|)$ (as explained in section 5.2.1)

H = Propagated hop count of the current packet for which route is being calculated

γ = Weight factor (0.5 in our case)

γ is to be adjusted in such a way that initially diverse paths will be selected but the progressively shortest hop route will get preference over the η-driven route to ensure convergence. When H and η are zero, α is used to find out the shortest path. Dijkstra's shortest-path algorithm has been modified to select a path having the smallest link weight (i.e., total link weight of all the links on that selected path will be minimal).

5.3.2 Finding the Maximally Zone-Disjoint Shortest Path: An Analysis

Let us assume that a packet, after propagating through H hops from source node, has arrived at an intermediate node n_i and it has to go to destination D. Let us also assume that the packet from n_i has two choices to reach destination D: a longer path (P_L) with low η and a shorter path (P_S) with high η. Let us also assume that the packet needs to traverse through h hops along P_S and through (h + x) hops along P_L to reach destination.

If our strategy were to find out the maximally zone-disjoint path, then obviously the longer path with low η would have been the choice. However, as discussed earlier, that would not be the optimal solution for improved throughput, as packets may get diverted toward longer paths unnecessarily, increasing the end-to-end delay. In the following analysis, we are trying to estimate a strategy for selecting the maximally zone-disjoint shortest path.

As mentioned before, link cost (n_i, n_j) for the current communication with communication-ID c = α + $η^{n_i}_c$ + γH:

So, sum of all link cost on path P_L = α × (h + x) + $η_c(P_L)$ + γ × H × (h + x),

Where, $η_c(P_L)$ = Correlation factor η of path P_L for communication-ID c (def. 8 and 9)

Similarly, cost of P_S = α × h + $η_c(P_S)$ + γ × H × h

The longer path P_L will be selected if cost of P_L < cost of P_S,

i.e., if (α × (h + x) + $η_c(P_L)$ + γ × H × (h + x)) < (α × h + $η_c(P_S)$ + γ × H × h)

Case I. If $η_c(P_L)$, $η_c(P_S)$ and H are zero (initial condition), then P_L will never get selected

Case II. If $η_c(P_L)$ = $η_c(P_S)$, then also P_L will never get selected. This implies that if correlation factors of two paths were the same, the shorter path would be selected.

Case III. If $\eta_c(P_L) <> \eta_c(P_S)$, then the longer path would be selected if $(\eta_c(P_L) + \gamma \times H \times (h + x)) < (\eta_c(P_S) + \gamma \times H \times h)$ [ignoring $\alpha \times x$, because $\alpha << \eta_c$], or $(\eta_c(P_S) - \eta_c(P_L)) > \gamma \times H \times x$

This implies that if the correlation factor of the shorter path is more such that the difference in the correlation factor of shorter path and that of longer path is greater than $\gamma \times H \times x$, then the longer path will be selected.

Termination Condition

If H >10, then the shortest path is selected irrespective of the value of η. If h = 1 (i.e., next hop is destination), then the destination is selected.

Example 1. Let us assume that the longer path is two hops longer than the shorter path (x = 2) and $\eta_c(P_S) = 5$ and $\eta_c(P_L) = 0$. So, the longer path is totally zone-disjoint from all other active paths whereas the shorter path is having $\eta_c(P_S) = 5$. From our above discussion, it follows that the longer path will be selected if $5 > 2 \times \gamma \times H$.

Case 1.1	H=2, $\gamma < 1.25$	→ Selects longer path
Case 1.2	H=2, $\gamma >= 1.25$	→ Selects shorter path
Case 2.1	H=8, $\gamma < 0.3125$	→ Selects longer path
Case 2.2	H=8, $\gamma >= 0.3125$	→ Selects shorter path

Example 2. Let us assume that the longer path is two hops longer than the shorter path (x = 2) and $\eta_c(P_S) = 9$ and $\eta_c(P_L) = 0$. So, the longer path is totally zone-disjoint from all other active paths whereas the shorter path is having high correlation factor $\eta_c(P_S) = 9$. The longer path will be selected if $9 > 2 \times \gamma \times H$.

Case 3.1	H=8, $\gamma < 0.5625$	→ Selects longer path
Case 3.2	H=8, $\gamma >= 0.5625$	→ Selects shorter path

In our simulation, we have set γ to 0.5. This implies that, as in case 1, when propagated hop count is two, i.e., the packet has traveled only two hops from source node to reach n_i, the longer path is selected in order to give a preference to low η. However, as in case 2, if the propagated hop count is eight, i.e., the packet has already traveled eight hops to reach n_i from source node, the shorter path is selected in order to give a preference to shortest path. At the same time, as shown in case 3, if the shorter path has a very high correlation factor

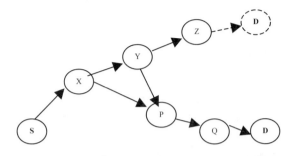

Figure 5.4 Adaptive route selection by intermediate node to reach destination D.

(= 9), then even if the propagated hop count is eight, the longer path is selected.

So, by adjusting the value of γ, we can adjust the preference between zone-disjointness and shortness of path. If γ is low, a packet would tend to take long, bypass low-η routes, whereas if γ is high, a packet would tend to take shorter routes. So, γ is termed the dispersion factor. Through several experimentations under different conditions, we have seen that the optimal value of γ is 0.5.

5.3.3 Adaptive Route Selection

Because the mechanism does not guarantee that each node would know the exact status of the network, an intermediate node corrects the routing decision and takes an alternative path to route data packets toward the destination. However, a node closer to the destination will have more accurate information about the destination and communication status in its neighborhood. This is illustrated in Figure 5.4. Source S has initially determined an approximate route S–X–Y–Z–D to reach D, where the dotted circle shows the initial position of D. However, due to mobility, node D changes its position and the current position of node D is shown with a solid line. As soon as this information of change of location of node D reaches the intermediate node Y, it decides to correct the path to node D because it has more accurate information about node D and can determine a better path toward node D through P and Q. Thus, path is getting selected and modified adaptively depending on the accuracy of available information, without generating a lot of control packets. Because each node has a GLST and ANL, this will improve the routing performance.

5.4 Maximally Zone-Disjoint Multipath Routing

5.4.1 Multipath Routing Using Omni-Directional and Directional Antennas

The routing schemes for ad hoc networks usually employ single-path routing. However, once a set of paths between source s and destination d is discovered, in some cases, it is possible to improve end-to-end delay by splitting the total volume of data into separate blocks and sending the blocks via selected multiple paths from s to d, which would eventually reduce congestion and end-to-end delay.[16] Utilization of multiple paths to provide improved performance, as compared to a single-path communication, has been explored in the past in the context of wired networks.[17,18] The application of multipath techniques in mobile ad hoc networks seems natural, as multipath routing allows diminishing the effect of unreliable wireless links and the constantly changing topology.[19] Moreover, it may also help to reduce end-to-end delay and perform load balancing. An on-demand multipath routing scheme is presented by Nasipuri and Das[20] as a multipath extension of Dynamic Source Routing (DSR),[21] in which alternate routes are maintained, so that they can be utilized when the primary one fails. It has been shown that the frequency of searching for new routes is much lower if a node keeps multiple paths to the destination. However, the performance improvement of multipath routing on the network load balancing has not been studied extensively. Perlman et al.[2] demonstrate that the multipath routing can balance network loads in their recent paper. However, their work is based on multiple channel networks, which are contention-free but may not be available in most cases. The Split Multipath Routing (SMR), proposed by Lee and Gerla,[6] focuses on building and maintaining maximally disjoint multiple paths.

However, it has also been shown that deployment of multiple paths does not necessarily result in a lower end-to-end delay. The effect of Alternate Path Routing (APR) in mobile ad hoc networks has been explored.[2] It was argued that the network topology and channel characteristics (e.g., route coupling) could severely limit the gain offered by APR strategies.

In this section, we have used the notion of zone-disjoint routes in wireless media and investigated the effect of directional antennas on multipath routing and compared their effectiveness with respect to multipath routing using omni-directional antennas. It has been shown that multipath routing is the best choice when the number of communications

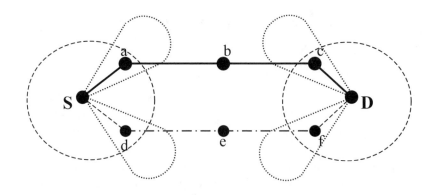

Figure 5.5 Two node-disjoint paths.

is low. However, with an increasing number of communications, single-path adaptive routing performs better than multipath adaptive routing. If the number of simultaneous communications is C and each communication uses two paths in multipath routing, then the number of traffic flows in multipath routing becomes 2C. So, as C increases, the number of traffic flows increases, creating more congestion and collision. As a result, the performance of multipath routing degrades as compared to single-path routing, when C is high.

In Figure 5.5, S and D are communicating using two paths: S–a–b–c–D and S–d–e–f–D. Because both a and d are within omni-directional transmission range of S as shown in the figure with a dotted circle, an RTS from S to a will also disable d. Similarly, because both c and f are within omni-directional transmission range of D, a CTS from D will disable both c and f. So, the lowest possible η in the case of omni-directional antennas [η_{min}(omni)] with two multipaths between s and d is two if two paths have no common nodes excepting source and destination. It is not possible to get fully zone-disjoint routes using omni-directional antennas. With directional antennas, it is possible to de-couple these two routes, making them fully zone-disjoint. If each node sets their transmission zone toward its target node only as shown in Figure 5.5, then the communication between S–a–b–c–D will not affect the communication between S–d–e–f–D. Hence, η_{min}(omni) = 2 whereas η_{min}(dir) = 0.

As a result, even if we get multiple zone-disjoint routes with minimal correlation factor using omni-directional antennas, the best-case packet arrival rate at the destination node will be one packet at every $2 \times t_p$, where t_p is the average delay per hop per packet of a traffic stream on the path p. The best-case assumption is single traffic stream in the

Table 5.3 Packet Arrival Rate at D with an Omni-Directional Antenna

	S	a	b	c	d	e	f	D
T_0	P_1>a							
T_1		P_1>b						
T_2	P_2>d		P_1>c					
T_3					P_1>D	P_2>e		P_1
T_4	P_3>a						P_2>f	
T_5		P_3>b					P_2>D	P_2
T_6	P_4>d		P_3>c					
T_7	...	...	...	...	...	...	...	...

network from S to D with error-free transmission of packets. Let us refer to Figure 5.5 and assume that each node is equipped with an omni-directional antenna and two paths having minimal correlation factor (i.e., η = 2) are used for data communication. This implies that nodes {a, b, c} and nodes {d, e, f} are disjoint.

Let us denote t_p as a time-tick, and at each time-tick a packet is getting transmitted from one node to other. Consider Table 5.3. S is sending data-packet P_1 to node a at time-tick T_0 and node a is forwarding data-packet P_1 to node b in the next time-tick (i.e., T_1). With an omni-directional antenna, S has to sit idle during T_1, because S has received an RTS from a. So, S can only transmit its second packet P2 to d (the first node of the second path) at time-tick T_2. From the packet transition schedule of S–D communication shown in Table 5.3 it is evident that destination D will receive packets in alternate time-ticks. Even if we increase the number of paths between s and d beyond two, the situation will not improve with omni-directional antennas.

In contrast, if we use directional antennas, the best-case packet arrival rate at destination will be one packet at every t_p.

Table 5.4 illustrates this point. With directional antennas, when node a is transmitting a packet to node b, S can transmit a packet to node d simultaneously. Thus, as shown in Table 5.4, destination D will receive a packet at every time-tick with two zone-disjoint paths using a directional antenna. Two zone-disjoint paths with directional antennas are sufficient to achieve this best-case scenario.

Initially we developed our own simulator to study the performance of multipath routing with omni-directional and directional antennas.

Table 5.4 Packet Arrival Rate at D with a Directional Antenna

	S	a	b	c	d	e	f	D
T_0	$P_1>a$							
T_1	$P_2>d$	$P_1>b$						
T_2	$P_3>a$		$P_1>c$		$P_2>e$			
T_3	$P_4>d$	$P_3>b$		$P_1>D$		$P_2>f$		P_1
T_4	$P_5>a$		$P_3>c$		$P_4>e$		$P_2>D$	P_2
T_5	$P_6>d$	$P_5>b$		$P_3>D$		$P_4>f$		P_3
T_6	$P_7>a$		$P_5>c$		$P_6>a$		$P_4>D$	P_4
T_7	...	...	...	...	...	...	...	...

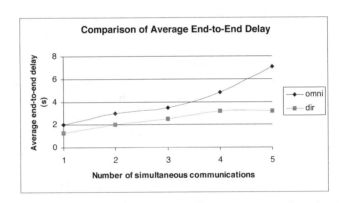

Figure 5.6 Increase in average end-to-end delay with multiple multipath communications using omni-directional and directional antennas.

There we have used ideal directional beam patterns in an environment of 40 nodes and observed that with an increasing number of simultaneous communications, the average end-to-end delay per packet increases much more sharply with omni-directional antennas compared to that with directional antennas as shown in Figure 5.6. So, it can be concluded that the routing performance using multiple paths improves substantially with directional antennas compared to that with omni-directional antennas. This is a consequence of reduced route coupling with directional antennas.

5.4.2 Selecting Maximally Zone-Disjoint Multipath Routes

The proposed multipath routing scheme selects a pair of maximally zone-disjoint shortest paths for each communication. Moreover, each pair of such paths also tries to avoid the nodes that are already handling several other communications. This way our multipath routing scheme is capable of balancing the network load, which in turn reduces the possibility of power depletion of some heavily used nodes in the network as well as the probability of congestion.

Each node in the network uses its current network status information (approximate topology information and ongoing communication information) to calculate the suitable next hop for reaching a specified destination via multiple paths such that the interference with the nodes that are already involved in some communication gets minimized. Our goal is to distribute the network load along a set of diverse paths to achieve better load balancing through multipath routing for an effective gain in throughput. In our proposed routing mechanism, each source node in the network is trying to distribute the data packets alternately along two maximally zone-disjoint paths. Each node consults its own active node list (ANL) to calculate zone-disjoint paths, so that the nodes that are already handling multiple communications (nodes with high correlation factor) may be avoided as far as possible in the current route selection process. As stated earlier in the context of single-path routing, here also we use correlation factor and propagated hop count to evaluate a route suitable for communication as well as to ensure the convergence of the selected multipath route. The link-weight calculation process, giving initial preference to zone-disjointness and subsequently giving preference to hop count, is the same as described in the context of a single-path routing scheme.

Initially, in the case of multipath routing, when a packet is to be transmitted by the source node for a communication-ID c, it is assigned a sub-ID cs_1 and the source S consults its ANL and GLST, assigns link weights, and selects a suitable next hop toward the destination to forward the packet along the least-cost path. Communication sub-ID cs_1 and the current next hop are kept as history information in source S for future route calculation. So, when the next packet comes to S for the same communication-ID c, a new sub-ID will be assigned to the packet, say cs_2, so that the packet may be routed along a path that is zone-disjoint compared to the earlier path selected by S. For the second packet, S tries to avoid the earlier zone (containing the next hop of the first packet) and calculates the next hop for the second packet in similar fashion. Then the history is overwritten with the new next hop and communication sub-ID cs_2. Next time, the communication sub-ID assigned to a third packet with communication-ID c will be cs_1. So sub-IDs will toggle between cs_1 and cs_2, which ensures the selection of alternate zone-disjoint shortest routes by the source node S. Basically, the source will transmit data packets along two zone-disjoint paths alternately.

Each intermediate node will adaptively select a suitable next hop toward the destination according to their ANL and GLST and assign suitable link weights, but will not keep any history information as source node. Because the mechanism does not guarantee that each node would know the exact status of the network, each node n in a path will compute its best next hop to reach the destination.

5.5 Performance Evaluation

5.5.1 Simulation Environment

The simulations are conducted using QualNet 3.1.[22] We have implemented the MAC protocol as illustrated in Chapter 3 and the routing protocol as illustrated before in the QualNet simulator with an ESPAR antenna and ideal directional antenna.

One hundred nodes are randomly placed over 1500 × 1500 square meters. Nodes are randomly chosen to be CBR (constant bit rate) sources, each of which generates 512 bytes of data packets to a randomly chosen destination at a rate of 2 to 500 packets per second. The set of parameters used is listed in Table 5.5.

5.5.2 Impact of Overhead

Because both GLST and ANL are periodic update packets and their propagations are limited to one-hop broadcasts, the network would never get flooded with ANL or GLST, as shown in the following analysis. In fact, we rely on approximate global network status information and accurate local status information similar to the fisheye concept.[11] So, an intermediate node adaptively modifies routing decisions based on more accurate local information around that node.

Let us assume that each update packet migrates at a time gap of T milliseconds and takes t milliseconds to physically migrate from one node to another. Let us also assume that our bounded region of ad hoc operation is A square meters, N is the number of nodes within

Table 5.5 Parameters Used in Simulation

Parameters	Value
Area	1500 x 1500 sq. m
Number of nodes	100
Transmission power	10 dBm
Receiving threshold	–81.0 dBm
Packet size	512 bytes
CBR packet arrival interval	2 ms to 500 ms
ANL periodicity (T_A)	500 msec
GLST periodicity (T_G)	10 seconds

A, and the omni-directional transmission range of each node is R. When a node is broadcasting an update packet to its neighbors, the nodes within the circular transmission zone around that node are busy, but nodes in other regions of area A can broadcast packets. Thus, in an average case, where the topology is evenly distributed over the region A, the number of zones in area A in which update packets could migrate between nodes simultaneously, without mutual interference, equals $(A/(\pi R^2))$. Now because the nodes are evenly distributed, the number of nodes (and consequently the number of update packets P) confined in a zone will be:

$$P = \frac{N}{A/(\pi R^2)} = \frac{N(\pi R^2)}{A}$$

In other words, P is the number of update packets that has to migrate from one node to another sequentially. As each update packet migrates at a time gap of T milliseconds and takes t milliseconds to do so, the medium will be occupied by update traffic [t.P × 100 / T] percent of the time.

In our case, the bounded region of operation is 1500 × 1500 square meters and R is 300 m, T_{GIST} is 10 seconds, and T_{ANL} is 500 msec. However, depending on the number of nodes, the size of a GLST update packet varies and it may be necessary to fragment one update packet into smaller packets for broadcasting purpose, considering the

Table 5.6 Overhead Analysis (Theoretical Values)

Number of Nodes	t_{ANL}	t_{GLST}	Total Overhead (%) [GLST+ANL] (Theoretical)
60	2 msec	9 msec	3.69264
80	3 msec	18 msec	7.83744
100	6 msec	54 msec	21.8544

maximum packet size of 1024 bytes. As we have seen through experimentation, a typical GLST update packet of a network of 100 nodes would require six fragmented packets to broadcast one GLST update packet whereas if the number of nodes is 60 or less, one packet is sufficient to broadcast one GLST update packet. If time to broadcast one packet is 9 msec, then one GLST update packet would take 9 × 6, that is 54 msec, to reach the next node. For ANL, the maximum size of an ANL update packet is less than 1024 bytes even for 100 nodes, so no fragmentation is required. At the same time, for fewer nodes, ANL packet length is less and consequently the time to broadcast ANL is less.

Simulation for overhead analysis has been performed on QualNet with static routes on a static topology of 100 nodes where nodes are randomly placed. The use of a static route is to study the performance with and without control traffic overhead irrespective of the routing strategy. The results of both the theoretical analysis (Table 5.6) and simulation study (Figure 5.7) show that the impact of overhead due to update packets is not at all significant for a number of nodes 60 or less. However, with an increasing number of nodes, increase in overhead is significant. However, in spite of this overhead, the performance improvement compared to that of AODV is always significant. Figure 5.8 shows that the experimental overhead is consistent with the calculated overhead. With an increasing number of nodes, the deviation of experimental results with calculated results increases due to more interference in the medium.

Control overhead is an integral part of any operational ad hoc network. Several researchers feel that on-demand, reactive routing schemes that do not use periodic messages of any kind would be more suitable in the context of ad hoc networks.[23,24] However, it has been observed that these protocols perform well under light traffic and low mobility, but performance degrades significantly under high mobility and high traffic load.[24,25] As mobility increases, the precomputed route may break down, requiring multiple route discoveries on

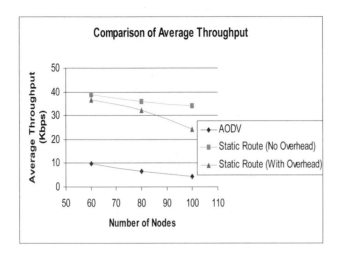

Figure 5.7   Impact of overhead on average throughput.

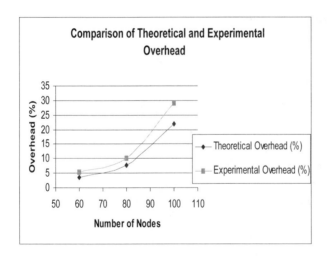

Figure 5.8   Comparison of theoretical and experimental overhead.

the way to destination. Route caching becomes ineffective in high mobility. Flooding is used for query dissemination and route maintenance, therefore on-demand routing tends to become inefficient due to control overhead when the frequency of communication requirement is high. On the other hand, several researchers have proposed proactive routing schemes based on classical distance vector or link-state routing.[23] In link-state routing, global network topology information is maintained in all routers by the periodic flooding of link-state updates by each node. Any link change triggers an immediate update. Unfortunately, as these schemes rely on flooding of routing updates, excessive control overhead may be generated, especially in a highly mobile environment. Thus, in the context of ad hoc networks, researchers have focused on restricting the propagation of routing updates, thereby reducing the control overheads. For example, Fisheye State Routing (FSR)[11] introduces the notion of a multilevel fisheye scope to reduce routing update overhead in large networks.

Our scheme is similar to the fisheye concept and tries to acquire approximate network status information through periodic propagation of update packets. This is the real gain in this scheme and serves as the single major motivation to replace the conventional link-state routing. The generation of control packets is fixed and does not depend on the mobility or number of simultaneous communications. It has been observed that, in conventional methods, the generation of control packets grows drastically with increase in mobility or number of communications.[25]

To summarize, the control overhead in our scheme is acceptable and is comparable, if not better, with other conventional schemes.

5.5.3 Evaluation under Static Scenarios

We have used AODV with IEEE 802.11 as its MAC as a benchmark to compare and evaluate the performance of our proposal. We evaluated ACR-single-path, ACR-multipath, and conventional shortest-path routing, all with ESPAR antennas, and we compared the throughput with AODV that uses an omni-directional antenna. Initially, we took a number of static snapshots and observed the performance, as compared to AODV. The average of our observations is shown in Figures 5.9 and 5.10 under the following scenarios:

- At a different number of simultaneous communications with a CBR packet arrival rate of 200 packets per second
- At a CBR packet arrival rate of 2 packets per second to 500 packets per second with packet size 512 bytes in a scenario of six communications

At a data rate of 200 packets per second, we have shown the throughput variation with an increasing number of simultaneous communications. When the number of communications is low, the issue of route coupling is much less significant, so shortest-path (Shortest-ESPAR)

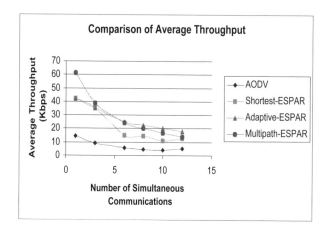

Figure 5.9 Comparison of average throughput with increasing number of simultaneous communications.

and ACR-single-path (Adaptive-ESPAR) performances are the same. However, with an increasing number of communications, the performance of shortest-path routing degrades faster than that of ACR-single-path. When the number of communications is ten, the ACR-single-path throughput is approximately double than that of shortest path.

When the number of communications is low, ACR-multipath (Multipath-ESPAR) outperforms other schemes. This is because, when the number of communications is low, a multipath scheme easily finds zone-disjoint paths for mutipath traffic flow for a particular communication and hence the performance improves. However, as the number of communications increases, say to four, then the multipath scheme would generate eight flows and under the present setting of 100 nodes in 1500 × 1500 square meters with a 300-meter transmission range, it becomes difficult to get so many zone-disjoint paths. So, the more flows there are, the greater the contention and congestion. As a result, with an increasing number of communications, ACR-multipath performance degrades as compared to that of ACR-single-path.

With an increasing number of communications, AODV performance suffers because of low SDMA (Space Division Multiple Access) efficiency due to the use of omni-directional antennas and route coupling. Therefore, the throughput of AODV is consistently low. ACR-single-path performance is three times (for lesser number of communications) to five times (for higher number of communications) more than that of AODV.

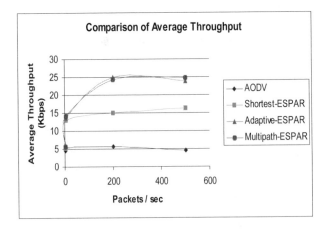

Figure 5.10 Comparison of average throughput with different packet arrival rate.

With multiple source destinations communicating at a time at a high data rate, the utilization of the medium can be increased to a large extent using directional antennas. Along with this, if we select maximally zone-disjoint paths, this will further reduce the contention among routes for getting access to the medium they share and we can get a scenario where the network load is balanced across all the nodes in the network. The combined effect of these two aspects will eventually improve the system performance drastically with improved throughput, as shown in Figure 5.10.

We have also evaluated the performance of ESPAR antennas as compared to ideal directional antennas. Figures 5.11 and 5.12 show the performance of ACR-single-path with an ESPAR antenna (Adaptive-ESPAR) vis-à-vis ACR-single-path with an ideal directional antenna (Adaptive-IDEAL). The results show that the performance of an ESPAR antenna is comparable to that of an ideal directional antenna.

5.5.4 Evaluation under Mobile Scenarios

Figure 5.13 shows the impact of mobility on ACR protocols and compares its performance with AODV. With mobility, the performance degradation of an ACR scheme is much less significant as compared to that of AODV. Because an ACR scheme relies on table-driven, adaptive routing, there is no impact of route failure: intermediate nodes adaptively correct the initial routing decision to take care of route failures. On the other hand, performance of AODV is heavily dependent on the

A Routing Strategy for Effective Load Balancing ■ 131

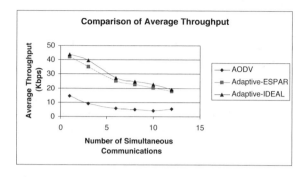

Figure 5.11 Performance of an ESPAR antenna compared to an ideal antenna with increasing number of simultaneous communications.

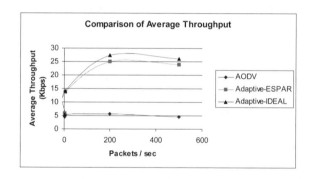

Figure 5.12 Performance of an ESPAR antenna compared to an ideal antenna with different packet arrival rates.

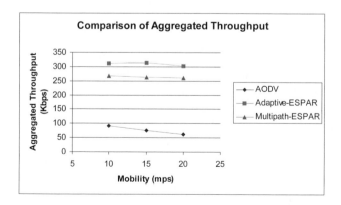

Figure 5.13 Comparison of aggregated throughput with increasing mobility.

number of route failures. As mobility increases, route errors due to route failures would increase, degrading the performance of AODV. The only impact of mobility on ACR is that, with increasing mobility, the network status information at any node tends to be less accurate. As a result, a packet may take a longer route to reach its destination because of adaptive route corrections at the intermediate nodes (Figure 5.4). As mentioned earlier, single-path routing performance is little better than multipath, because the number of simultaneous communications is six.

5.6 Discussion

Use of directional antennas in ad hoc wireless networks can drastically improve system performance, if we consider the issue of routing with load balancing along with suitable directional MAC protocols. Maximally zone-disjoint shortest routes will be helpful in this context to reduce route coupling among selected paths and thereby to improve throughput. In spite of the control overhead incurred due to periodic propagation of GLST and ANL in the network, the performance is far better than conventional reactive routing with omni-directional MAC protocols. Our table-driven adaptive routing with maximally zone-disjointness exploits the advantages of a directional antenna and improves system performance.

It has been shown that multipath routing may not improve system performance: if the number of communications is high where each communication is using multiple paths, the performance is worse than single-path. Multipath routing is the best choice when the number of communications is low. However, with an increasing number of communications, single-path adaptive routing performs better than multipath adaptive routing.

Based on this observation, future work could be to investigate the feasibility of priority-based multipath routing where one or two high-priority nodes will route data using multipath routing, whereas other nodes will use single-path routing. This may improve the throughput of communications with higher priority.

Another issue for future work is scalability. It may be possible to adjust the periodicity of GLST propagation adaptively. For example, if the node density is high, then all the nodes in a neighborhood need not broadcast GLST, as it will not improve the perception of a node about the network status to a large extent. Similarly, if the average system mobility is low, GLST periodicity may be increased in an adaptive fashion. However, detection of node density or system mobility in a dynamic environment would require additional control packets. In any case, the performance of this scheme is always better than conventional routing schemes with an omni-directional antenna, especially when the system mobility is high or the number of simultaneous communications is high.

References

1. Wu, K. and Harms, J., On-demand Multipath Routing for Mobile Ad Hoc Networks, in *Proc. EPMCC 2001*, Vienna, February 20–22, 2001.
2. Pearlman, M.R. et al., On the Impact of Alternate Path Routing for Load Balancing in Mobile Ad Hoc Networks, in *Proc. MobiHOC 2000*, p. 150.
3. Bandyopadhyay, S. et al., Multipath Routing in Ad Hoc Networks with Directional Antenna, in *Proc IFIP TC6/WG6.8 Conference on Personal Wireless Communications (PWC 2002)*, Singapore, October 2002.
4. Hassanein, H. and Zhou, A., Routing with Load Balancing in Wireless Ad Hoc Networks, in *Proc. 4th ACM International Workshop on Modeling, Analysis and Simulation of Wireless and Mobile Systems*, Rome, Italy, 2001.
5. Lee, S.J. and Gerla, M., Dynamic Load-aware Routing in Ad Hoc Networks, in *Proc. IEEE ICC 2001*, Helsinki, Finland, June 2001.
6. Lee, S.J. and Gerla, M., Split Multi-path Routing with Maximally Disjoint Paths in Ad Hoc Networks, in *Proc. ICC*, 2001.
7. Bandyopadhyay, S. et al., Multipath Routing in Ad Hoc Wireless Networks with Omni Directional and Directional Antenna: A Comparative Study, in *Proc. International Workshop on Distributed Computing*, Calcutta, December 27–30, 2003. Lecture notes in Computer Science (LNCS 2571), Springer-Verlag, December 2002.
8. Das, S.K. et al., An Adaptive Framework for QoS Routing through Multiple Paths in Ad Hoc Wireless Networks, in *Journal of Parallel and Distributed Computing (JPDC)* Vol. 63, No. 2, 2003, p. 141.
9. Parkins, C.E. and Royer, E.M., Ad Hoc On Demand Distance Vector Routing, in *Proc. 2nd IEEE Workshop on Mobile Comp. Sys. and Apps.*, February 1999, p. 90.
10. Parkins, C.E. et al., Ad Hoc On Demand Distance Vector (AODV) Routing, http://www.ietf.org/internet-drats/draft-ietf-manet-aodv-03.txt, June 1999. IETF Internet draft.
11. Pei, G., Gerla, M., and Chen, T.W., Fisheye State Routing: A Routing Scheme for Ad Hoc Wireless Networks, in *Proc. IEEE International Conference on Communication*, New Orleans, Louisiana, June 2000.

12. Roy Choudhury, R., Bandyopadhyay, S., and Paul, K., A Distributed Mechanism for Topology Discovery in Ad Hoc Wireless Networks Using Mobile Agents, in *Proc. of the First Annual Workshop On Mobile Ad Hoc Networking and Computing (MOBIHOC 2000)*, Boston, Massachusetts, August 11, 2000.
13. Roy Choudhury, R., Paul, K., and Bandyopadhyay, S., An Agent-based Connection Management Protocol for Ad Hoc Wireless Networks, in *Journal of Network and System Management*, December 2002.
14. Nasipuri, A. et al., On-demand Routing Using Directional Antennas in Mobile Ad Hoc Networks, in *Proc. IEEE International Conference on Computer Communication and Networks (ICCCN2000)*, Las Vegas, Nevada, October 2000.
15. Roy Choudhury, R. and Vaidya, N., Impact of Directional Antennas on Ad Hoc Routing, in *Proc. 8th Conference on Personal and Wireless Communication (PWC)*, Venice, September 2003.
16. Das, S.K. et al., Improving Quality-of-Service in Ad Hoc Wireless Networks with Adaptive Multi-path Routing, in *Proc. GLOBECOM 2000*, San Francisco, California, November 2000.
17. Rao, N.S.V. and Batsell, S.G., QoS Routing via Multiple Paths Using Bandwidth Reservation, in *Proc. IEEE INFOCOM*, 1998.
18. Bahk, S. and El-Zarki, W., Dynamic Multi-path Routing and How It Compares with Other Dynamic Routing Algorithms for High Speed Wide-area Networks, in *Proc. ACM SIGCOM*, 1992.
19. Tsirigos, A., Haas, Z.J., and Tabrizi, S.S., Multi-path Routing in Mobile Ad Hoc Networks or How to Route in the Presence of Frequent Topology Changes, in *Proc. MILCOM*, 2001.
20. Nasipuri, A. and Das, S.R., On-demand Multipath Routing for Mobile Ad Hoc Networks, in *Proc. IEEE ICCCN*, Boston, Massachusetts, October 1999.
21. Johnson, B. and Maltz, D.A., Dynamic Source Routing in Ad Hoc Wireless Networks, in *Mobile Computing*, T. Imielinski and H. Korth, Eds., Kluwer, Boston, 1996.
22. QualNet Simulator Version 3.1, Scalable Network Technologies, www.scalable-networks.com.
23. Royer, E.M. and Toh, C.K., A Review of Current Routing Protocols for Ad Hoc Wireless Networks, in *Proc. IEEE Personal Communication*, April 1999, p. 46.
24. Das, S.R., Perkins, C., and Royer, E., Performance Comparison of Two On-demand Routing Protocols for Ad Hoc Networks, in *Proc. IEEE INFOCOM 2000*, Tel Aviv, March 26–30, 2000.
25. Paul, K., Roy Choudhury, R., and Bandyopadhyay, S., Survivability Analysis Ad Hoc Wireless Network Architecture, in *Proc. Second International Workshop on Mobile and Wireless Communication Networks*, Paris, France. Lecture Notes in *Computer Science*, LNCS 1818, Springer-Verlag, May 2000.

Chapter 6

Priority-Based QoS Routing Protocols Using Smart Antennas

6.1 Introduction

With the expanding possibility of applications of ad hoc wireless networks, the need for supporting quality of service (QoS) in these networks is becoming essential. Numerous solutions to the QoS problems have been proposed in this context.[1-7] However, these protocols did not consider a major aspect of wireless environments, i.e., mutual interference. Interference between nodes within the proximity of each other causes route coupling.[8-10] This can be avoided by using zone-disjoint routes, as illustrated in Chapter 5.

In this chapter, our primary objective is to propose a priority-based routing scheme, which will protect the high-priority flows from the contention caused by the low-priority flows. Our protocol avoids the coupling between routes used by high- and low-priority traffic by reserving a high-priority zone of communication. The part of the network, used for high-priority data communication, will be temporarily reserved as the high-priority zone. Low-priority flows will try to avoid this zone by selecting routes that are maximally zone-disjoint[10] with respect to the high-priority reserved zone and will consequently reduce the contention between high- and low-priority flows in that reserved zone. But, this does not ensure that the low-priority flows will be able to avoid the high-priority zone completely. As the number of high-priority flows increases in the network, it becomes difficult for the low-priority flows to find routes avoiding high-priority zones. Some topological situation may also occur where some low-priority flows may not get a path through any unreserved part of the network. As a result, low-priority flows will be forced to take routes through a high-priority zone, causing interference. This may be controlled by temporarily blocking such low-priority flows in the system. Low-priority flows will constantly monitor the reservation status of the network in order to find a path through an unreserved zone. As soon as a low-priority flow gets such a path, either due to mobility of nodes or the end of a high-priority session, it immediately resumes the blocked communication. In this chapter, we will also discuss the effectiveness of low-priority call blocking to improve the throughput of high-priority flows in a network consisting of several coupled high- and low-priority flows.

Intervehicular communication (IVC) on highways is one of the major application areas in ad hoc networks that enable multi-hop data exchange and forwarding.[11] While driving, there is a constant need for local information regarding roadblocks, traffic conditions, and any accidents ahead. Also, in medical emergencies, information about nearby hospitals or the availability of doctors in nearby cars may be obtained through multi-hop data exchange and forwarding. So, these emergency applications may require some messages to be forwarded on a top-priority basis to the intended destination that brings forward the issue of service differentiation among the data flows in the network. Service differentiation ensures that flows belonging to a higher "service class" must receive better service.[12] This application area of an ad hoc network has motivated us to explore the possibility of applying our priority-based routing strategy in this scenario. We have identified some interesting issues related to routing in this kind of unbounded network and proposed a priority-based *reactive* routing mechanism using directional antennas to achieve service differentiation in this kind of environment. This scheme ensures a timely and reliable delivery of high-priority data by minimizing the effect of interference caused by low-priority flows in a high-priority communication zone.

Several researchers have already studied and addressed the issues related to QoS in ad hoc networks. QoS support in the context of ad hoc networks includes QoS models, QoS resource reservation signaling,

QoS routing, and Media Access Control.[1-7] However, Pallot and Miller[7] have proposed that limited bandwidth of the mobile radio channel prevents giving every class of traffic the same QoS except when the network is very lightly loaded. So, some means for providing each class a different QoS must be implemented by assigning priority to one class over another class in terms of allocating resources. Thus, linkage between QoS and priority is a common one in the literature, and the two terms are almost interchangeable.[7] So, QoS provisioning through priority-based service is an interesting idea that is worth exploring.

Several efforts have also been made to support QoS in ad hoc networks by changing the size of the contention window (CW) according to the priority of traffic in the MAC layer and modifying the back-off algorithm accordingly.[13] Because this approach is probabilistic, it does not guarantee that high-priority packets will always get a contention-free access to the medium for data communication. Moreover, two high-priority flows contending for the medium may not always get guaranteed fair access of the medium in these schemes.

None of the existing priority-based QoS routing protocols[1-7] consider the effect of mutual interference between routes in wireless media during routing, though coupling is one of the major causes for degradation in network performance. Both end-to-end delay and throughput, the two major concerns in supporting priority-based QoS in ad hoc networks, are heavily dependent on route coupling. To get rid of the effect of coupling between routes during data communication, the notion of zone-disjoint routes was proposed in Chapter 5. If zone-disjoint routes are used, then data communication along one path will not interfere with data communication along other paths and simultaneous communications will be possible.

Our objective is to exploit the advantage of zone-disjointness and use it to calculate diverse routes for low-priority flows, which will minimally interfere with a zone containing high-priority traffic. But getting zone-disjoint or even partially zone-disjoint paths using an omni-directional antenna is difficult because the transmission zone of an omni-directional antenna covers all directions. A directional antenna has a narrower transmission beamwidth compared to an omni-directional antenna. So, two interfering routes can be easily decoupled using a directional antenna.[10] In this chapter, we have proposed to use a directional antenna in our priority-based routing using zone reservation.

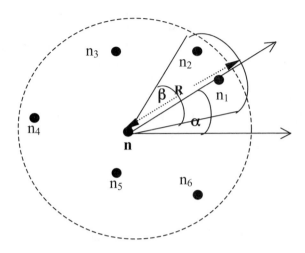

Figure 6.1 Transmission zone$_n$(α,β,R) and omni-directional transmission range (in dotted lines) showing directional and omni-directional neighbors.

6.2 A Few Related Definitions

In this section, we will discuss the key terms related to our proposal and will subsequently illustrate the basic mechanism to find zone-disjoint routes to avoid route coupling in wireless media:

- Zone: When a node n forms a transmission beam at an angle α and a beamwidth β with a transmission range R, the coverage area of n at an angle α is defined as transmission_zone$_n$ (α,β,R) (Figure 6.1) of node n. Because transmission beamwidth β and transmission range R are fixed in our study, we will refer to transmission_zone$_n$ (α,β,R) as transmission_zone$_n$ (α) or zone of commuication$_n$ (α) or simply zone$_n$ (α) in subsequent discussions. The nodes lying within transmission_zone$_n$ (α) are known as the directional neighbors of n at an angle α. Hence, only n_1 and n_2 are directional neighbors of n at an angle α in Figure 6.1.
- High-priority zone: It is the transmission_zone$_n$ (α) formed by any node n that is involved in high-priority communication. If n→n_1 is an ongoing high-priority communication, then transmission_zone$_n$ (α), shown in Figure 6.1, is the high-priority zone. The directional neighbors of n at an angle α (n_1 and n_2) are then known as reserved directional neighbors as they are reserved for high-priority communication, n→n_1.

- Zone reservation: To reserve the zone at a node n at an angle α for a communication n→n_1 in Figure 6.1, the status of node n and the status of each directional neighbor of n at an angle α are set as reserved. Thus, zone reservation essentially sets the status of all directional neighbors of a node at a particular beam pattern including that node as reserved so that other communications may avoid those reserved nodes during their route calculation process. Avoiding the reserved zone of a communication actually eliminates the possibility of interference caused by other communications to that ongoing communication. In our proposed protocol, zones reserved by high-priority flows are avoided by low-priority flows during their route selection process.
- Reserved node list (RNLn): This contains the perception of node n about high-priority communication activities in the entire network. As mentioned earlier, it is a set of nodes at an instant of time t where each node is either a sender or a receiver in any high-priority communication process or a directional neighbor of this sender node. Each node in this list is associated with a set of communication-IDs for which it is reserved.

6.3 Priority-Based QoS Routing Using Zone Reservation

Our protocol assigns a path to a high-priority flow that is the shortest as well as maximally zone-disjoint with respect to other high-priority communications. Each low-priority flow will try to take an adaptive zone-disjoint path avoiding all high-priority zones. If such a path is not available, it will block the flow adaptively to protect high-priority flows. Thus, for low-priority flows, a shortest path criterion is not a predominant metric. However, unless we consider the hop count or path length of low-priority flows, packets belonging to low-priority flows may get diverted toward longer paths unnecessarily, increasing the end-to-end delay. Moreover, there is no assurance of convergence: the packets may move around the network in search of zone-disjoint paths but may not reach the destination at all. Thus, our proposal ensures the following:

- High-priority traffic will be unaffected by low-priority traffic. As a result, in a network having multiple flows with different priorities, the high-priority throughput will remain almost the

same as if the network is handling only high-priority flows. If more than one high-priority flow is present in the network, then each will try to take a path that will be maximally zone-disjoint with respect to other high-priority flows.
- Low-priority flows will try to find paths that are zone-disjoint with respect to high-priority flows and at the same time, the total hop count of each path is less than a predefined maximum path length LP_{maxhop}. If such a path is not available, then the low-priority call is blocked temporarily.

6.3.1 Zone Reservation and Route Computation by High-Priority Flows

Each node in the network uses its current network status information (approximate topology information and ongoing high-priority communication information) to calculate the suitable next hop for reaching a specified destination such that the interference with reserved nodes gets minimized. Initially, when a packet is transmitted from a source, it gives preference to the zone-disjoint path-selection criteria. If a packet reaches an intermediate node after traversing multiple hops, then a progressively shorter hop route toward the destination will be selected. So this adaptive route calculation mechanism guarantees the convergence of the proposed routing algorithm. We have used the following function to calculate the link weights that will ensure the selection of a lower η path for a low propagated hop count and the selection of a lower hop path for a higher propagated hop count. Dijkstra's shortest-path algorithm has been modified to select a path having the smallest link weight, i.e., the total link weight of all the links on that selected path will be minimum.

Link cost (ni ,nj) during the current communication having communication-ID $c = \alpha + \eta^{ni}_c + \gamma H$ where,

α = Initial link weight (.01 in our case; $\alpha \ll \eta_c$ and $\alpha \ll H$)
η^{ni}_c = The sum of the total number of communications (excepting the current communication c handled by each reserved directional neighbor in the directional zone (n_i->n_j)
H = Propagated hop count of the current packet for which route is being calculated. Propagated hop count indicates the number of hops already traversed by a packet at any point of time
γ = Dispersion factor

Priority-Based QoS Routing Protocols Using Smart Antennas

By adjusting the value of γ, we can adjust the preference between zone-disjointness and the shortness of the path. If γ is low, a packet would tend to take long, bypass low-η routes, whereas if γ is high, a packet would tend to take shorter routes. So, γ is termed the dispersion factor. Through several experiments under different conditions, we have seen that the optimal value of γ is 0.5 for high-priority flows.

Each flow in the network is identified with a unique ID and belongs to either a high- or low-priority category. Whenever a packet sent by a high-priority flow comes to a node for a particular destination, the node simply selects the lowest-cost path toward that destination and transmits the packet to the immediate next hop on the selected path. The lowest-cost path for high-priority flow is calculated as follows, according to the formula shown above:

- Each link (n_i, n_j) in the network is initialized with a constant value $(= \alpha)$.
- If a pair of nodes n_i and n_j are involved in some high-priority communication, then all the nodes in the directional transmission zone of the sender n_i toward receiver n_j will set their activity status as high to indicate that they should sit idle to support high-priority transmission. They are treated as reserved nodes and are updated in RNL. RNL of a node is transmitted periodically to all other nodes so that they can update their RNLs.
- Link weight of each link connected to a reserved node (nodes belong to RNL) is set to a value based on the formula described above. Accordingly, when a source of a high-priority flow calculates its route, our path-selection algorithm will automatically selects the maximally zone-disjoint shortest path.
- If a reserved node does not receive a high-priority packet for a considerable period of time, then it will set itself as unreserved so that other communications may select paths through it, if required.

6.3.2 Route Computation and Adaptive Call Blocking by Low-Priority Flows

6.3.2.1 Route Computation without Call Blocking

When call blocking is not used, the low-priority flows try to select longer, diverse routes to avoid the high-priority zone as far as possible using the notion of maximally zone-disjointness. Whenever a packet sent by a low-priority source comes to a node for a particular destination, the node simply selects the lowest-cost path toward that destination and transmits the packet to the immediate next hop on the selected path. The lowest-cost path for low-priority flow is calculated using the same formula described above with dispersion factor $\gamma = 0.2$. A low value of dispersion factor implies that the selected route will be longer than a route selected with a high value of dispersion factor. That means low-priority traffic will select a longer but diverse route to avoid a high-priority zone as far as possible whereas high-priority traffic will select a shorter, diverse route with respect to other high-priority routes to reduce interference among multiple high-priority flows. But this does not ensure that the low-priority flows will be able to avoid the high-priority zone completely. As the number of high-priority communications in the network increases, more nodes will get involved in high-priority communication within the high-priority reserved zone. Thus, low-priority flows may not get enough unreserved nodes to route their traffic without disturbing high-priority flows. Sometimes, the current network topology and communication pattern of the coupled flows may not allow a low-priority flow to take any alternative route excepting a route through a high-priority zone even if there is enough unreserved area in the network (see Figure 6.4 later in this chapter). As a result, low-priority flows will not be able to find a path avoiding reserved nodes. In these cases, low-priority flows will be forced to go through high-priority zone, unless we use call blocking.

6.3.2.2 Route Computation with Call Blocking

Here, low-priority flows will try to find a path that is totally zone-disjoint with respect to high-priority flows and at the same time, the total hop count of the path is less than a predefined maximum path length LP_{maxhop}. Let us assume that a packet belonging to a low-priority flow, after propagating through H hops from source node, has arrived at an intermediate node n_i and it has to go to destination D. Let us also assume that the packet needs to traverse h more hops to reach a destination along a path that is totally zone-disjoint with respect to high-priority flows to ensure convergence of low-priority packet delivery at the destination (H + h) should be less than LP_{maxhop}.

If such a path is not available, the low-priority flows will block themselves temporarily to protect high-priority performance. Low-priority call blocking is implemented as follows:

- A low-priority source will consult its RNL and select a zone-disjoint route avoiding reserved nodes. At the same time, the total hop count of such a route should be less than a predefined maximum path length LP_{maxhop} (18 hops in our case).
- If no such route is available, the low-priority communication will be stopped temporarily.
- When the next low-priority packet is to be transmitted, the node will try to find out a suitable route toward the destination again as before. The absence of high-priority flows in a high-priority reserved zone for a long time automatically sets the status of a reserved node as unreserved and the reserved node list (RNL) at each node gets updated periodically with current status information. So, it is possible that the low-priority flow will be able to find a suitable route through an unreserved part of the network now and will be able to resume the blocked communication. Thus, the low-priority flows in our scheme are adaptively blocking and resuming the communication as per the demand of the situation to protect high-priority flows.

6.3.3 Performance Evaluation

The proposed routing protocol is implemented on QualNet simulator using an ESPAR antenna with 12 overlapping patterns at 30-degree intervals[14,15] as our directional antenna pattern to prove the effectiveness of our proposal. The simulation environment specifications and parameters used are described in Table 6.1.

Initially we tried to establish that zone reservation is an effective means to provide priority-based QoS in an ad hoc wireless network. For this, we have selected six random source-destination pairs (Flow1 to Flow6) as illustrated in Section 6.3.3.1. But zone-reservation protocol alone does not work well in some scenarios where low-priority flows do not get any suitable unreserved zone for routing and are forced to take a route through the high-priority reserved zone. This in turn affects the performance of high-priority flows. In such a situation adaptive call blocking of low-priority flows is a necessity. To establish this fact, we have chosen random source-destination pairs in such a way that

Table 6.1 Parameters Used in Simulation

Parameters	Value
Area	1500 x 1500 sq. m
Number of nodes	100
Transmission power	10 dBm
Receiving threshold	–81.0 dBm
Sensing threshold	–81.0 dBm
Data rate	2 Mbps
Packet size	512 bytes
Simulation time	5 minutes
Mobility model	Random way-point
Packet injection rate	15 ms
Topology	Random

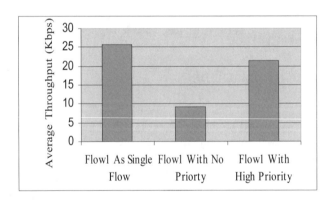

Figure 6.2 Behavior of a particular flow (Flow1) with different priority assignments in a scenario of six communications.

low-priority flows do not get any suitable unreserved zone for routing. This is illustrated in Section 6.3.3.2.

6.3.3.1 Effectiveness of Zone-Reservation Protocol

We have considered the following scenarios and initially observed the throughput of Flow1 (Figure 6.2) when:

1. Only Flow1 is present in the network and selects the shortest path (shown in Figure 6.2 as "Flow1 As Single Flow").
2. Flow1 is communicating in the presence of five other flows (Flow2–Flow6) in the network and no priority scheme is used. Thus, all of them select shortest paths that may be interfering with each other (shown in Figure 6.2 as "Flow1 With No Priority").
3. A priority-based service differentiation scheme is employed. Flow1 is assigned high priority and thus takes the shortest path. Moreover, Flow1 reserves a directional zone around each node on its route so that five other low-priority flows will eventually select adaptive paths, avoiding the zone reserved by Flow1. The throughput of Flow1 in this scenario is shown in Figure 6.2 as "Flow1 With High Priority."

In Figure 6.2, in the first case, throughput of Flow1 is maximum, which is an obvious outcome of the fact that no other flow is causing any disturbance to it. In the second case, as all the flows are using shortest paths, the existence of route coupling among those routes reduces the throughput of Flow1 drastically. But, in the third case, as soon as high priority is assigned to Flow1 and routes are selected according to our protocol, throughput of Flow1 shows a remarkable improvement, which is almost the same as the throughput in the first case.

We have also observed (in Figure 6.3) the average throughput of the five low-priority flows (i.e., Flow2–Flow6) under the situation described above. Figure 6.3 illustrates that if high priority is assigned to Flow1 then the average throughput of five low-priority communications reduces a little bit in comparison to the corresponding average throughput when no priority is assigned to Flow1. The low-priority flows avoid high-priority zones and select routes away from this zone. So, low-priority flows may have to travel longer hops to reach their destinations but will experience less interference. This explains the small reduction in throughput of average low-priority flows in Figure 6.3.

6.3.3.2 Effectiveness of the Call-Blocking Scheme

Figure 6.4 depicts a typical scenario where call blocking is the only way to protect a high-priority flow from low-priority interference. The high-priority zone (HP Zone) in Figure 6.4 is reserved in such a way

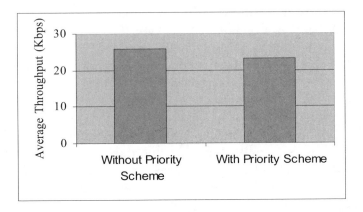

Figure 6.3 Behavior of five low-priority flows with and without assigning priority to one flow (Flow1) in a scenario of six communications.

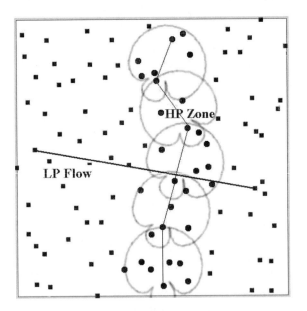

Figure 6.4 A typical scenario where call blocking is necessary.

that the low-priority flow (LP Flow) will not be able to find any path avoiding that zone. So, the low-priority flow has no other alternative but to take a path through the HP zone. As a result high-priority throughput will suffer due to interference caused by the low-priority flow. Thus, it is intuitive that blocking the low-priority flow may be

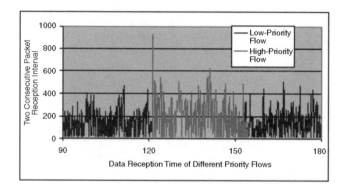

Figure 6.5 Adaptive call blocking of a low-priority flow in the presence of a high-priority flow.

an effective means to handle such a situation. The effectiveness of the call-blocking scheme is shown in earlier figures, assuming this kind of scenario in a network consisting of several coupled high- and low-priority flows.

In Figure 6.5 the packet reception interval between two consecutive packets of each flow in a scenario of two coupled flows (one high- and one low-priority) throughout the simulation period is shown. In this scenario, the low-priority flow starts at 90 seconds and ends at 180 seconds, whereas the high-priority flow starts at 120 seconds and continues up to 150 seconds.

We are trying to show the effect of adaptive call blocking of a low-priority flow when the low-priority flow cannot find any unreserved zone from source to destination to route its packets. As the RNL is propagated to all the nodes in the network, the nodes become aware of the ongoing high-priority flow and the low-priority flow starts to block itself, sensing the nonavailability of a suitable path and avoiding the reserved zone. When the high-priority source stops injecting packets after 150 seconds, then the low-priority source again starts its communication. It is clear from this graph that only high-priority flow operates during the period of 120 to 150 seconds.

Single HP with Mutually Coupled Single LP Flow

Initially we have chosen two coupled flows and assigned high priority to one of them. The routes are chosen in such a manner that the low-priority flow has no other option but to go through some portion of

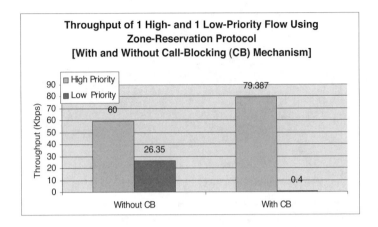

Figure 6.6 Effect of call-blocking mechanism on the throughput of one high-priority flow and one low-priority flow coupled with each other.

the reserved zone of a high-priority flow. The performance of high- and low-priority flows are shown in Figure 6.6. The performances of high- as well as low-priority flows are compared before and after the implementation of a call-blocking mechanism. It is observed that high-priority throughput is improving substantially by blocking the low-priority flow, which was creating contention to the high-priority flow. Blocking the low-priority flow reduces contention in the medium so that the high-priority flow can get full access to the medium, which results in improved throughput for the high-priority flow. The massive degradation of low-priority throughput is due to call blocking. Because we have taken snapshots of static scenarios, where the low-priority flow is not getting any alternative path to avoid the high-priority reserved zone, the low-priority flow has to block itself. Because it will take some time to take this blocking decision (it depends on the propagation time of RNL to source node), the throughput of the low-priority flow under call blocking is not zero, but close to zero.

It is very likely that mobility of nodes will change this snapshot and the low-priority flow will get an alternate path through an unreserved zone to adaptively unblock itself.

Multiple HP Flows with Multiple LP Flows

Static Scenario — Figure 6.7 illustrates the advantage of a zone-reservation protocol with call blocking as compared to a simple zone-reservation protocol (without any call-blocking scheme). The scenario

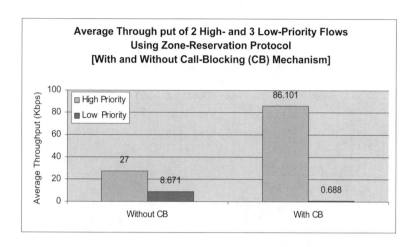

Figure 6.7 Effect of call blocking on the average throughput of two high-priority flows and three low-priority flows coupled with each other.

chosen is the average of a set of static settings with two high- and three low-priority flows where all are coupled flows.

As before, degradation of low-priority throughput is due to call blocking. Because low-priority flows are not getting any alternative paths avoiding the high-priority reserved zones, they have to block themselves. Because it will take some time to take this blocking decision, the average throughput of low-priority flows under call blocking is not zero, but close to zero.

Mobile Scenario — We have also evaluated the above scheme of low-priority call blocking along with a zone-reservation protocol under moderate mobility (0–10 mps) and the throughput performances of both high- and low-priority flows are shown in Figure 6.8. Average throughput of high-priority flows is showing a phenomenal improvement using a low-priority call-blocking technique compared to the corresponding throughput in a simple zone-reservation protocol without a call-blocking technique. Moreover, degradation of low-priority throughput in the mobility scenario is much less compared to the static scenario because the low-priority flows that block themselves without getting any paths avoiding the high-priority reserved zone will soon be able to select alternate paths through the unreserved area in the network and release the blocked calls, as mobility of nodes adaptively changes the paths selected by high-priority flows and the corresponding high-priority reserved zone also changes. So under mobility, it is

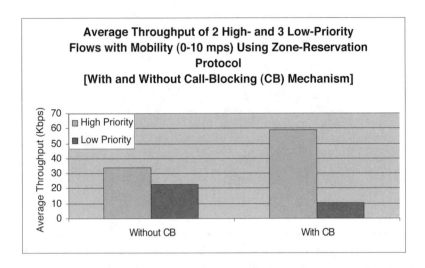

Figure 6.8 Effect of call-blocking mechanism on the average throughput of two high-priority and three low-priority flows coupled with each other under mobility (0–10 mps) using zone reservation.

possible to adaptively block and unblock the low-priority flows at a faster rate depending on the mobility of nodes.

6.4 Priority-Based Flow-Rate Control for QoS Provisioning Using Feedback Control

Several researchers have explored the idea of a control-theoretic approach for flow rates in the context of a wired network to control congestion in the network and to provide flow-based end-to-end QoS as well as to deal with fairness issues.[18–21] Kolarov and Ramamurthy[18] proposed a control mechanism that can be used to design a controller to support available bit rate service, where users would dynamically share the available bandwidth in an equitable fashion, by adjusting an appropriate set of distributed controls based on feedback of explicit rates. Keshav[19] presented a control-theoretic approach that can be used to control transport connections in reservationless networks. These flow control schemes have used feedback control mechanisms to allow the flows present in the network to share the available bandwidth equally among them depending on the network status at that instance. But these schemes have not addressed any sort of priority-based service

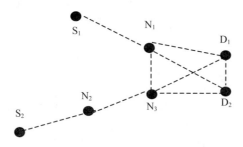

Figure 6.9 Low-priority flow (S_2–D_2) is disturbing high-priority flow (S_1–D_1) because of route coupling. Dotted lines show omni-directional connectivity among nodes.

issues to give preference to some flows over the others. Moreover, they have investigated the wired network environment.

Two flows in an ad hoc wireless network will affect each other when the two routes belonging to these two different flows share common nodes, or when they are close enough to interfere with each other, causing route coupling.[8] In this case, nodes in those two routes will constantly contend for access to the medium they share. This is shown in Figure 6.9. In such a situation, if the flow rate of the low-priority flow is reduced, the high-priority flow will get more chances to access the medium they share, which eventually reduces the congestion and improves the throughput of the high-priority flow. Thus, priority-based flow control is an effective means to provide service differentiation to different classes of flows.

Researchers have talked about end-to-end flow control in the transport layer to achieve service differentiation.[12] But these schemes cannot ensure desired rates for high-priority traffic. Earlier work on service differentiation through rate control of a flow[12] focuses on individualized flow control. According to that scheme, flows are controlled individually with a rate vector based on end-to-end feedback, where high-priority flows are throttled less aggressively than low-priority flows. So, the high-priority flow rate will improve but may not be able to maintain a desired level of flow.

Our objective is to adaptively maximize low-priority flows while maintaining high-priority flows at a desired level so that full utilization of the wireless medium is achieved through adaptive rate control. To provide this desired service differentiation to high-priority flows, we need a flow-control algorithm, where the low-priority flows, causing interference to a high-priority flow, detect and measure the high-priority flow rate at each node on their routes and consequently adjust their flow rates using a control-theoretic approach to protect the high-priority flow at its desired level. This detection and measurement is done at the MAC layer of each node participating in communication. This would enable the nodes involved in the low-priority flow (e.g., S_2, N_2, N_3, D_2 in Figure 6.9) to measure the high-priority flow rate in the vicinity by recording consecutive RTS reception intervals from nodes involved in high-priority flows (e.g., N_3 will receive RTSs from N_1 involved in the high-priority flow and can measure the flow rate of the high-priority flow at each instant of time). This information is conveyed to S_2, the source node of the low-priority flow, which will compute the control decision and adjust its packet injection rate adaptively to maintain the high-priority flow rate at its desired level.

To implement the scheme, we have proposed to use special types of RTS and CTS packets. There would be an extra field in the RTS packet, which denotes the communication-ID and priority level of the flow to which the packet belongs. This extra field in RTS is required to make the neighbors aware of the priority level of the ongoing communication. Similarly, the CTS packet also has two extra fields. The first field is similar to the extra field of the RTS packet and is required to convey the priority level of the ongoing communication to its neighbors. The second field contains the maximum packet-arrival-interval of the high-priority communication in its vicinity. Even in the presence of more than one high-priority flow in the neighborhood of a low-priority flow, back propagation of the maximum packet arrival interval of the high-priority flows is done. This indicates that the low-priority flow can adaptively adjust itself repeatedly, so that the high-priority flows can get maximum chances to the medium and their expected packet arrival intervals are maintained.

For example, in Figure 6.9, let us assume that there was a continuous low-priority flow S_2–N_2–N_3–D_2. When operating alone, its flow rate is fixed at a predefined value. Now, a high-priority flow S_1–N_1–D_1 starts. Let us assume that we want to fix and maintain this high-priority flow rate at a predefined level. However, because these two routes are close enough to cause route coupling, they will interfere with each other, which will reduce the flow rate of the high-priority flow at the interfering nodes N_1 and D_1. Our objective is to detect this reduced flow rate of the high-priority flow at nodes belonging to the low-priority flow and back-propagate this knowledge back to the low-priority source, which then can adaptively reduce its flow rate to maintain the high-priority flow rate at its predefined value. To implement this, from the RTS transmitted by N_1 and CTS transmitted by D_1, both N_3 and D_2 detect the high-priority flow S_1–D_1. This remains unknown to the source S_2, which is far away from the high-priority flow. So, with the help of the CTS packet, D_2 transmits the knowledge to N_3. When N_3 has to send a CTS packet to N_2, it combines its own detection of high-priority traffic with the received knowledge from D_2 and cumulatively considers the contention in the flow and transmits it with the CTS packet. N_2 lastly sends this information back to S_2 with a CTS packet. The source node, S_2, then considers the contention in the medium of the flow and adaptively makes a decision to reduce packet injection rate. Hence, with no extra packet, the information of contention in the medium of a high-priority flow is transmitted to the low-priority source node, which adaptively reduces the packet injection rate. So, when there is no contention in the medium, even a low-priority flow can operate at its predefined flow rate.

The above mechanism is different from other existing MAC-layer solutions for service differentiation.[22-24] Several efforts have been made to support QoS in MANET by changing inter frame spaces (IFS) and the size of the contention window (CW) according to the priority of traffic in the MAC layer and by modifying the back-off algorithm accordingly. But it does not guarantee that the high-priority packet will always get contention-free access to the medium for data communication.[24] Multiple high-priority flows contending for the medium may not always get guaranteed fair access of the medium in these schemes. Moreover, multiple low-priority traffic in the absence of high-priority traffic may choose a large contention window leading to poor utilization of the medium. Another important aspect of QoS in the MAC layer, which has not been addressed by the researchers, is the packet delivery ratio. Low-priority packets in the MAC layer of intermediate nodes may often choose an increased back-off counter that remains unknown to the source node, which may still be injecting packets at a very high rate. As a consequence, the packets arriving at a very high rate at intermediate nodes, handling low-priority flow, are not served quickly by the MAC layer and remain in queue, which may overflow and cause packet drops.

What we try to achieve in our approach is a specified level of service guarantee in terms of flow rate to high-priority flows when they contend with low-priority flows. We have proposed this protocol with a very nominal overhead using an omni-directional antenna and modified the scheme to show the overall improvement in throughput using a directional antenna.

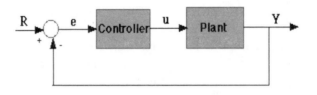

Figure 6.10 Basic feedback controller.

6.4.1 A Control-Theoretic Approach

6.4.1.1 Some Preliminaries on Proportional-Integral-Derivative (PID) Control

A feedback controller (Figure 6.10) is designed to generate an output u that causes some corrective effort to be applied to a process so as to drive a measurable process variable Y toward a desired value R known as the set point. The controller uses an actuator to affect the process and a sensor to measure the results. Virtually all feedback controllers determine their output by observing the error e between the set point (R) and a measurement of the process variable (Y). Errors occur when a disturbance or a load on the process changes the process variable. The controller's mission is to eliminate the error automatically.[25]

Earlier feedback control devices implicitly or explicitly used the idea of proportional, integral, and derivative (PID) actions in their control structure. However, it was Minorsky's phenomenal work on ship steering published in 1922 where rigorous theoretical consideration was given to PID control.[26] PID controller is still the most widely used control structure in modern industrial processes.[27]

The general form of the PID control algorithm is:

$$u = K_p e + K_i \int e\, dt + K_d \frac{de}{dt}$$

The variable (e) represents the tracking error, the difference between the desired input value is (R) and the actual output (Y). This error signal (e) will be sent to the PID controller, and the controller computes both the derivative and the integral of this error signal. The signal (u) just past the controller is now equal to the proportional gain (K_p) times the magnitude of the error plus the integral gain (K_i) times the integral of the error plus the derivative gain (K_d) times the derivative of the error.

Proportional gain (K_p) will have the effect of reducing the rise time and will reduce, but never eliminate, the steady-state error. An integral gain (K_i) will have the effect of eliminating the steady-state error, but it may make the transient response worse. A derivative gain (K_d) will have the effect of increasing the stability of the system, reducing the overshoot, and improving the transient response.

The above equation is a continuous representation of the controller and it must be converted to a discrete representation. There are several methods for doing this, the simplest being to use first-order finite differences.

The proportional error term will still use the error between the reference and the current value. The differential and integral terms can be replaced by:

$$\frac{de}{dt} = \frac{[e(n) - e(n-1)]}{\Delta t}$$

and

$$\int e(t)dt = \sum_{k=n-w}^{n} e(k) \times \Delta t$$

So the final form of the equation is:

$$m(n) = k_p \times e(n) + k_i \times \sum_{k=n-w}^{n} e(k) \times \Delta t + k_d \times \frac{[e(n) - e(n-1)]}{\Delta t}$$

Thus it will be necessary to find the current error, the sum of the errors, and the recent change in error to calculate desired output.

6.4.1.2 Priority-Based Flow Control Strategies Using a PID Controller

Figure 6.11 shows the basic flow-rate control scheme where a single high-priority and single low-priority flow are coupled with each other, i.e., the routes of these two flows either share a common node(s) or they are close enough to interfere with each other. Here, to maintain the high-priority flow at a desired flow rate R, the low-priority flow adaptively changes its flow rate u at its source using PID control strategy, so that the high-priority flow rate Y becomes as close to R

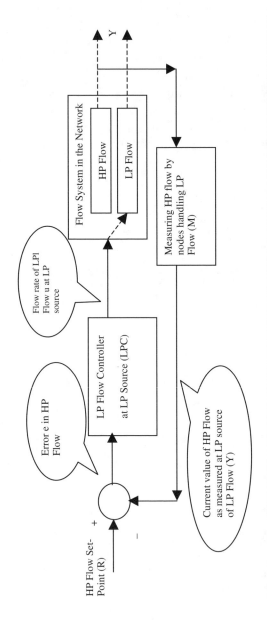

Figure 6.11 Basic flow-rate control scheme with a single high-priority and a single low-priority flow coupled with each other (HP = high priority; LP = low priority).

as possible. Each node handling low-priority flow is measuring high-priority flow, if that node is coupled with the high-priority flow. The block M, in Figure 6.11, conceptualizes this measurement. This information is back-propagated to the source S_L handling low-priority flow. From this, the node S_L determines Y and takes a control decision using PID control. Based on this, S_L regulates the flow rate of the low-priority flow and the process is repeated.

There is a subtle but important difference between the conventional PID controller as illustrated before and our proposed control scheme. It can be easily seen that in our control scheme, if we set u (i.e., the low-priority flow rate) to zero, Y will achieve its desired flow rate of R in the absence of any other low-priority flow. So, if the focus is only on maintaining the high-priority flow at a desired value, the solution does not require any controller in the conventional sense of the term. However, in the present context, our objective is to maximize the low-priority flow rate R_L as well, keeping the high-priority flow rate R_H at its desired level. This is similar to max-min flow control where our control strategy will maintain R_H at its desired level with a dynamically adjusted controlled setting of R_L in such a manner that R_L cannot be increased further without decreasing R_H from its desired value. This kind of requirement is absent in conventional PID control and, therefore, our approach is a derivative of conventional PID control, which we will illustrate subsequently.

Figure 6.12 shows the flow-rate control scheme where a single high-priority and multiple low-priority flows are coupled with one another. Here, to maintain the high-priority flow at a desired flow rate R, each low-priority flow adaptively changes its flow rate at its source using PID control strategy. Nodes handling each low-priority flow L_i are measuring high-priority flow (indicated by M_i) and this information is back-propagated to the source node of each low-priority flow. Based on this, each low-priority flow is regulated and the process is repeated.

Figure 6.13 shows the flow-rate control scheme where multiple high-priority and multiple low-priority flows are present and they are coupled. Here, the desired flow rate R assigned for a single high-priority flow is no longer valid, because coupling between two or more high-priority flows will not allow all of them to maintain that flow rate even in the absence of low-priority flows. So, in the case of multiple high-priority flows coupled with one another, a new set point for high-priority flows needs to be determined dynamically. For example, if the system allows a single high-priority flow to operate at a guaranteed flow rate of 50 packets/sec, the same system cannot guarantee the same flow rate in case two coupled high-priority flows are

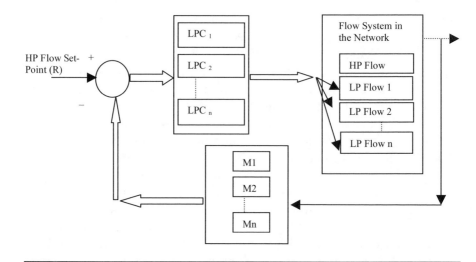

Figure 6.12 Flow-rate control scheme with single high-priority and multiple low-priority flows coupled with one another: each low-priority flow is using a similar control scheme to protect a high-priority flow.

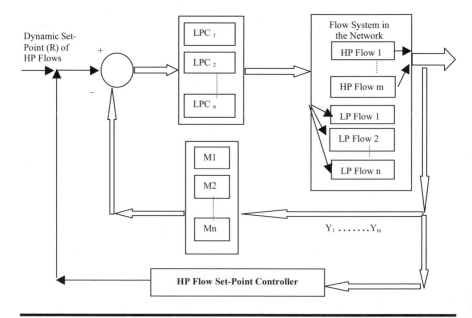

Figure 6.13 Flow-rate control scheme with multiple high-priority and multiple low-priority flow coupled with one another: each low-priority flow is using a similar control scheme to protect high-priority flows.

there in the system. In this case, the guaranteed flow rate for each of them will be, say, 30 packets/sec. And other low-priority flows will now try to protect that new flow rate assigned to high-priority flows. To change the set point of high-priority flows dynamically in a distributed environment, each high-priority flow monitors the existence of other high-priority flows in its neighborhood. If a high-priority flow H_1 detects one more high-priority flow H_2 in its neighborhood, implying that these two high-priority flows are coupled, H_1 will change its set point to a predefined value R_{new1} ($R > R_{new1}$ in packets per second). In a similar fashion, H_2 will also detect H_1 and will also change its set point to R_{new1}. If H_1 detects two high-priority flows in its neighborhood, H_1 will change its set point to another predefined value R_{new2}. These set points may be predefined experimentally or can be adaptively computed for optimal medium utilization. However, for simplicity, we have assumed that the set points are predefined experimentally. When H_1 detects the absence of any other high-priority flow in its neighborhood for a specified period of time, H_1 flow rate will revert back to R.

A low-priority flow will measure the high-priority flow rates in its neighborhood and will also get the set points of high-priority flows. Accordingly, it will adjust its flow rate to protect the *weakest* high-priority flow. Weakest high-priority flow in this context is a flow having the largest error value e = $(R_{new}-Y)$. This will naturally ensure the protection of other high-priority flows in its neighborhood.

6.4.2 Priority-Based Flow-Control Scheme Using Directional Antennas

6.4.2.1 Detecting and Measuring High-Priority Flow Rates by Other Flows

When a flow is initiated, packets are sent through multiple hops and at the MAC layer the packet delivery at each intermediate node is ensured by RTS-CTS-DATA-ACK exchange. These RTS and CTS packets are utilized to detect and back-propagate the flow-related information on which packet injection rate control decisions are made at low-priority sources.

As shown in Figure 6.14, a low-priority flow will interfere with a high-priority flow, only if the direction of flows overlaps. In Figure 6.14, although N_1 and N_3 are within the omni-directional transmission range of each other, the flow from N_1 to D_1 will not interfere with the flow from N_3 to D_2. To ensure that, it is imperative that each node in

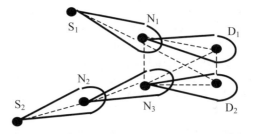

Figure 6.14 Using directional antenna low-priority flow (S_2–D_2) can coexist with high-priority flow (S_1–D_1).

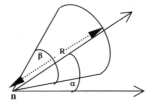

Figure 6.15 Transmission zone$_n(\alpha,\beta,R)$.

low-priority flow should sense whether its directional transmission zone in the direction of low-priority flow contains any node handling high-priority flow. If it does, it implies that the low-priority flow will disturb the high-priority flow and, consequently, it is necessary to control low-priority flow rate to protect high-priority flow rate. The mechanism for detecting and measuring high-priority flow rate by any node n is given below.

Definition 1

When a node n forms a transmission beam at an angle α and a beamwidth β with a transmission range R, the coverage area of n at an angle α is defined as transmission_zone$_n$ (α,β,R) (Figure 6.15) of node n. It implies that if a node m is within the transmission_zone$_n$ (α,β,R) and m is in receive mode, whenever n transmits a message at that transmission angle α with respect to n and beamwidth β and transmission range R, it will be received by m. When node m moves out of the transmission_zone$_n$ (α,β,R), the connectivity between n and m is lost. Because transmission beamwidth β and transmission range R are fixed here, we will refer to transmission_zone$_n$ (α,β,R) as transmission_zone$_n$ (α) in subsequent discussions.

Definition 2

$RRT^{Hi,\alpha,n}(t)$ (RTS-Reception-Time) is defined as the time t at which a node n receives an RTS at an angle α with respect to node n from any node currently handling the high-priority Flow H_i.

Definition 3

$PAI^{Hi,\alpha,n}(t)$ (packet arrival interval) is defined as the interval between two consecutive RRTs received at node n from the high-priority Flow H_i at an angle α with respect to node n at time t. This is used to measure the high-priority flow rate by any node n in its directional neighborhood as $(1/PAI^{Hi,\alpha,n}(t))$ packets/unit time. So,

$$PAI^{Hi,\alpha,n}(t) = RRT^{Hi,\alpha,n}(t) - RRT^{Hi,\alpha,n}(t_{previous}),$$

where $t-\Delta t < t_{previous} < t$ and Δt is the time-band introduced to ensure the validity of consecutiveness of two RTS packets arriving at node n. For example, if node n misses an RTS due to random channel error, collision, or mobility, it might wrongly calculate the flow rate. In this context, introduction of Δt is necessary. In the case of a high-priority destination node, which will never issue an RTS, CTS reception time is monitored to calculate the high-priority flow rate at the destination node of that high-priority flow.

Definition 4

The $PAIT^n(t)$ (PAI table) is defined as the packet arrival interval table at node n at time t which stores the $PAI^{Hi,\alpha,n}(t)$ for each high-priority Flow H_i at each angle α, as shown in Table 6.2.

Now, whether a low-priority flow at node n is creating contention with a high-priority flow depends on the transmission direction or transmission zone of the low-priority flow and ongoing high-priority communication in that zone. Let us assume that a low-priority flow at node n is using a transmission zone β with respect to node n. In other words, β is the direction of a low-priority flow at node n at an instant of time t. Whether this low-priority flow at node n is creating contention with any high-priority flows in its neighborhood depends on the PAIT entry at an angle β. That is, if PAIT at an angle β contains some PAI entry for high-priority flows as $\{<PAI^{Hj,\beta,n}(t)><PAI^{Hk,\beta,n}(t)>\ldots\ldots\}$, then $MAXPAI(\beta, t) = (MAX \{<PAI^{Hj,\beta,n}(t)>, <PAI^{Hk,\beta,n}(t)>\ldots\ldots\}$ needs to be back-propagated for rate adjustment of the low-priority flow.

Table 6.2 PAI Table Structure at Node n at Time t

Angle	Packet Arrival Interval of Each High-Priority Flow
α_1	{............}
α_2	{............}
...	{............}
α_i	{<$PAI^{Hj,\alpha i,n}(t_1)$> <$PAI^{Hk,\alpha i,n}(t_2)$>............} where t-Δt < t_p <= t
...	{............}
α_n	{............}

MAXPAI(β, t) will help the node handling low-priority flow to identify the most affected high-priority flow (as the maximum of these PAI values indicate minimum flow rate). Accordingly, the low-priority flow should try to adjust its flow rate to protect this maximum PAI, which in turn will automatically protect the other high-priority flows.

Let us now look into the distributed measurement technique of high-priority flow rate by the source node of a low-priority flow. If $S \to N_1 .... \to N_{k-1} \to N_k \to N_{k+1} \to .....D$ is a route from source to destination of a low-priority flow, then each node is independently measuring the flow rates of high-priority flows using $PAI^{Hi,\alpha,Nk}(t)$. The perceived flow rate of a high-priority flow at each node of a low-priority flow may be different. But, to take a control decision, the source node of a low-priority flow should be aware of the bottleneck flow-rate information, i.e., the lowest of all the measured value of flow rates at the nodes on the path $S \to N_1 .... \to N_{k-1} \to N_k \to N_{k+1} \to .....D$. In other words, the source node S should know the *maximum* of MAXPAI at all nodes on the path $S \to N_1 \to ... \to N_{k-1} \to N_k \to N_{k+1} \to .....D$ in the direction of the low-priority flow.

To simplify our discussion, so far we have considered only one low-priority flow. However, a node may be involved in multiple low-priority flows. So, MAXPAI needs to be associated with flow-ID and the direction of flow of that flow-ID, rather than the absolute value of any angle β. So, we modify the definition of MAXPAI as follows in definition 5.

Definition 5

$DMPAI(N_k)^{Li}(t)$ (detected maximum packet arrival interval) at node N_k handling the low-priority Flow L_i at time t is defined as the maximum

PAI of the high-priority flow in the direction of the low-priority flow L_i detected at the node N_k.

Definition 6

$PPAI(N_k)^{Li}(t)$ (propagated packet arrival interval) is the propagated value of maximum PAI of the high-priority flow at node N_k from node N_{k+1} as measured by the nodes $N_{k+1} \to \ldots D$ of the low-priority flow L_i. So:

$$PPAI(N_k)^{Li}(t) = MAX \{ DMPAI(N_{k+1})^{Li}(t) \ldots DMPAI(D)^{Li}(t) \}$$

$$PPAI(N_{k-1})^{Li}(t) = MAX \{ DMPAI(N_k)^{Li}(t), PPAI(N_k)^{Li}(t) \}$$

So, $PPAI(S)^{Li}$ is the maximum packet-arrival-interval of the high-priority flows, finally detected by the source node S of low-priority flow L_i. Source node S of L_i uses this information to adaptively control the flow rate of L_i.

6.4.2.2 Feedback Control of Low-Priority Flow Rate

We have considered the packet injection interval (PII) at the source node as a measure to controlling flow rate. The packet injection rate (PIR) of flow (in packets/sec) at a source node is computed at PII = 1/PIR. For example, if PII at a source node is 20 msec, then PIR = 50 packets/sec.

Let us now refer back to Section 6.4.1. To take any control decision, first we have to compute the error term in the PID controller.

Error e at any low-priority flow L_i at its source node S = (R − PPAI $(S)^{Li}$), where R is the desired high-priority PAI and $PPAI(S)^{Li}$ is the maximum packet-arrival-interval of high-priority flows in the neighborhood of L_i, detected by nodes in L_i and propagated back to the source node S of L_i. We assume that each high-priority flow has a prespecified packet injection interval R, which should correspond to the PAI at any intermediate node when high-priority flow does not have to face any contention. Every node in the network knows this value and this corresponds to the desired high-priority PAI. As discussed in Section 6.4.1.2, in the context of multiple high-priority flows, if a high-priority flow H_1 detects one more high-priority flow H_2 in its neighborhood, implying that these two high-priority flows are coupled, H_1 and H_2 will change their set points to a new predefined value R_{new} and this information will be conveyed to the low-priority flows in their neighborhood. When H_1 detects the absence of other high-priority flows in its neighborhood for a specified time period, H_1 flow rate will revert back to R.

Once the error $e(n)$ and the time interval between two successive error Δt is calculated, the PII of L_i (S) is calculated as:

$$\text{PII}(\text{new}) = \text{PII}(\text{old}) - \left[k_p \times e(n) + k_i \times \sum_{k=n-w}^{n} e(k) \times \Delta t + k_d \times \frac{[e(n) - e(n-1)]}{\Delta t} \right]$$

The values of k_p, k_i, and k_d need to be tuned for optimal performance. The performance of the controller is shown in the next section.

6.4.3 Performance Evaluation

We evaluated the performance of our proposed scheme on a QualNet simulator.[29] We considered IEEE 802.11–based directional MAC[14] and implemented the proposed protocol with a directional antenna only. We have simulated an ESPAR antenna[14,28] in the form of a quasi-switched-beam antenna, which is steered discretely at an angle of 30 degrees, covering a span of 360 degrees. We have done the necessary changes in the QualNet simulator to implement the proposed protocol. The set of parameters used are listed in Table 6.3.

Table 6.3 Parameters Used in Simulation

Parameters	Value
Transmission power	15 dBm
Receiving threshold	–81.0 dBm
Sensing threshold	–91.0 dBm
Data rate	2 Mbps
Packet size	512 bytes
Simulation time	5 minutes

6.4.3.1 Performance of Low-Priority-Flow-Controller (LPC)

For simplicity, we assume grid topology with six nodes per row, with two extreme nodes acting as source and destination, respectively, whenever required. We have evaluated the performance of LPC under three settings of the grid topology:

- Single high-priority flow along one row of the grid and single low-priority flow along the second row of the grid: both routes are physically close enough to cause route coupling.
- Single high-priority flow along the second row of the grid and two low-priority flows along the first and third rows of the grid: all three routes are physically close enough to cause route coupling.
- Two high-priority flows along the second and third rows of the grid and one low-priority flow along the first row of the grid: all are physically close enough to cause route coupling.

We have used static routes to avoid the effects of routing protocols and to illustrate clearly the gain obtained in our proposed protocol. Also, we have used static routes to stop all the control packets generated by any routing protocol. Instead of random selection of a source-destination pair, we have chosen the source-destination pair so that there is contention between high- and low-priority flows to artificially create a situation so that we can demonstrate the effect of packet injection interval control.

Performance of LPC with One High-Priority Flow and One Low-Priority Flow

Figures 6.16(a), 6.16(b), and 6.16(c) demonstrate the performance of LPC with one high-priority flow and one low-priority flow. The desired set point R of packet injection interval (PII) of the high-priority flow is 20 msec, i.e., 50 packets/sec. With a packet size of 512 bytes and 100 percent packet delivery ratio, the expected throughput at PII = 20 msec is 204.8 Kbps. The LPC behavior is shown in Figure 6.16(a).

Here, the initial value of both PII(H) and PII(L) is 20 msec. Because high-priority flow cannot sustain this flow rate with a coupled low-priority flow operating at PII(L) = 20 msec, the PII of low-priority flow immediately shoots up to protect the high-priority flow rate. Gradually, the low-priority flow rate settles down to an average PII(L) = 60 msec

Table 6.4 Results of Adaptive Packet Injection Rate Control

Parameters	High-Priority Flow (H)	Low-Priority Flow (L)
K_P, K_I, K_D	—	0.8, 0.1, 0.1
Packet delivery ratio	1.00	0.999144
Throughput (Kbps)	204.8	53.11147

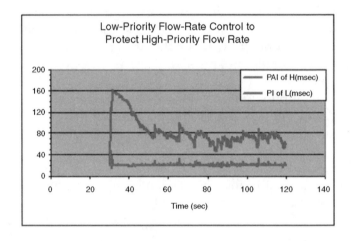

Figure 6.16(a) Adaptive flow-rate control of low-priority flow (L) to protect the high-priority flow.

(approximately). The corresponding throughput and packet delivery ratio is given in Table 6.4. High-priority flow is able to sustain a throughput of 204 Kbps with a packet delivery ratio of 1. The low-priority flow throughput is 53 Kbps with a packet delivery ratio of 0.999. The values of K_p, K_i, and K_d used in this result are 0.8, 0.1, and 0.1.

Figures 6.16(b) and 6.16(c) show the behavior of the system under the said condition without any controller. When both high- and low-priority flows are set at PII = 20 msec, throughput of the high-priority flow is 129 Kbps and that of low-priority flow is 124 Kbps with packet delivery ratio 0.63 and 0.6, respectively. We have measured the throughput and packet delivery ratio of the high-priority flow by varying manually the PII of the low-priority flow. By increasing the PII of the low-priority flow (PII(L)), both throughput and packet delivery ratio of the high-priority flow will increase. At PII(L) = 60 msec or more, we are getting peak throughput of the high-priority flow (204 Kbps) with a packet delivery ratio of 0.993. So, PII(L) = 60 msec is

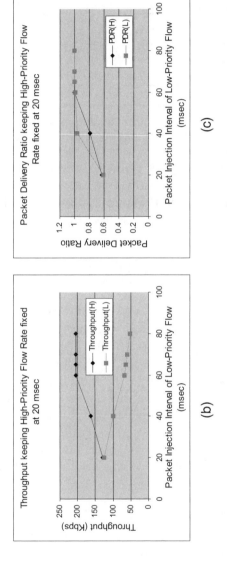

Figure 6.16(b) Throughput and (c) packet delivery ratio of one high- (H) and one low-priority flow (L), keeping the high-priority packet injection interval fixed at 20 msec (i.e., flow rate fixed at 50 packets/sec) with manually increasing packet injection interval of low-priority flow.

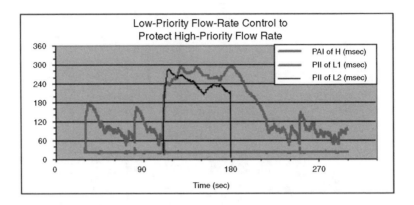

Figure 6.17 Adaptive flow-rate control of two low-priority flows (L1 and L2) to protect the high-priority flow (H) and maximize their own throughput.

the optimal, manually adjusted set point of the low-priority flow at which we can sustain the high-priority flow at its desired level. That is what we are achieving with the low-priority-flow-controller in Figure 6.16(a). Low-priority throughput is 67 Kbps at that set point.

Performance of LPC with One High-Priority Flow and Two Low-Priority Flows

Figure 6.17 shows the performance of LPC with one high-priority flow and two low-priority flows, all coupled with one another. The desired set point R of packet injection interval (PII) of the high-priority flow is 20 msec, i.e., 50 packets/sec. The LPC behavior is shown in Figure 6.17. Here, initial values of PII(H), PII(L_1), and PII(L_2) are 20 msec. To show the control action of LPC, H and L_1 starts simultaneously at 30 seconds and continues until the end of the simulation, whereas L_2 starts at 110 seconds and finishes at 180 seconds. In the absense of L_2, the behavior of LPC is the same as before. But when L_2 starts, PII of both L_1 and L_2 shoot up to protect the flow rate of H at PII(H) = 20 msec. Gradually, the low-priority flow rate settles down to an average PII(L) = 220 msec (approximately). After the withdrawal of L_2, PII of L_1 settles at its older value as shown. The corresponding throughput and packet delivery ratio is given in Table 6.5. High-priority flow is able to sustain a throughput of 204.8 Kbps as before with a packet delivery ratio of 1. The throughput of L_1 is 35.15 Kbps with a packet delivery ratio of 0.99 and that of L_2 is 17.5 Kbps with a packet delivery ratio of 1. The difference in throughput of L_1 and L_2 is because of the

Table 6.5 Results of Adaptive Packet Injection Rate Control with One High-Priority Flow and Two Low-Priority Flows

Parameters	Flows with Controller LPC_1 and LPC_2			Flows without Controller (All PII at 20 msec)		
	H	L_1	L_2	H	L_1	L_2
K_P, K_I, K_D	---	0.8, 0.5, 0.2	0.8, 0.5, 0.2	----	0,0,0	0,0,0
Packet delivery ratio	1.00	0.99	1.00	0.4	0.42	0.41
Throughput (Kbps)	204.8	35.15	17.5	82.06	86.42	83.29

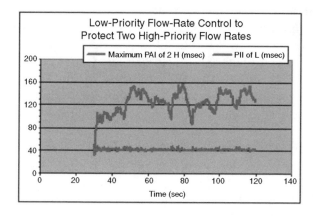

Figure 6.18 Adaptive flow control of a low-priority flow (L) to protect two high-priority flows and maximize its own throughput.

fact that L_1 operated at different PII(L_1) in the presence and absence of L_2. The values of K_p, K_i, and K_d used in this result are 0.8, 0.5, and 0.2. Without any controller, the throughput of H is 82.06 Kbps with a packet delivery ratio of 0.4.

Performance of LPC with Two High-Priority Flows and One Low-Priority Flow

Figure 6.18 shows the performance of LPC with two high-priority flows and one low-priority flow, all coupled with one another. The desired set point R of packet injection interval (PII) of the high-priority flow

Table 6.6 Results of Adaptive Packet Injection Rate Control with Two High-Priority Flows and One Low-Priority Flow

	Flows with Controller LPC			Flows without Controller (All PII at 20 msec)		
Parameters	H_1	H_2	L	H_1	H_2	L
K_P, K_I, K_D	—	—	0.8, 0.8, 0.1	—	—	0,0,0
Packet delivery ratio	0.998667	0.999111	0.998654	0.85	0.88	0.89
Throughput (Kbps)	102.2635	102.309	33.77	87.45	90.66	91.96

is now set to 40 msec, i.e., 25 packets/sec., because the system cannot sustain comfortably two high-priority flows at a lower PII value. This has been found out experimentally. Now the task of LPC is to maintain both high-priority flows H1 and H2 to this desired set point. The LPC behavior is shown in Figure 6.18. Here, initial values of PII(H_1), PII(H_2), and PII(L) are 40 msec. After initial shoot up, the low-priority flow settles down to an average PII(L) = 120 msec (approximately). The corresponding throughput and packet delivery ratio are given in Table 6.6. Both the high-priority flows are able to sustain a desired throughput of 102.3 Kbps with a packet delivery ratio of almost 1. Because PII (H) is now 40 msec, the ideal throughput with 100 percent packet delivery ratio with packet size 512 bytes is 102.4 Kbps. The throughput of L is 33.77 Kbps with a packet delivery ratio of almost 1. The values of K_p, K_i, and K_d used in this result are 0.8, 0.8, and 0.1.

6.4.3.2 Evaluating the System Performance in Random Topology

After evaluation of LPC in an ideal environment of grid topology without any overhead, we have evaluated the system performance in a random topology. We have implemented rotational sector–based receiver-oriented directional MAC protocol[14] with a location-tracking mechanism as our directional MAC protocol. We have used a network-aware routing protocol[15] as our directional routing protocol. On this MAC and routing protocol, we have implemented our proposed scheme to show the effectiveness of our proposal in a real scenario with overhead of various MAC and routing layer control packets. In a random topology of 100 nodes in a bounded region of 1500 × 1500 square meters, we have randomly chosen one source-destination pair as the high-priority flow and five more source-destination pairs as low-priority flows. We have evaluated the system performance in both a static scenario and under mobility.

Static Scenario

Figure 6.19(a) shows the performance of a high-priority flow in the presence of five low-priority flows with a flow-rate control scheme to compare its performance when (1) no LPCs are assigned to the low-priority flows and (2) the high-priority flow operates alone in the absence of any other flow. When the high-priority flow operates alone, it yields a throughput of nearly 86 Kbps. When five other flows are introduced and no flow-control scheme is assigned, the performance of this high-priority flow degrades to nearly half of its previous value, as evident from Figure 6.19(a). With the proposed flow-rate control scheme, throughput of the high-priority flow is almost the same as it was while operating alone.

Figure 6.19(b) shows the performance of five low-priority flows in two scenarios: without any LPC and with LPC. After introduction of the proposed scheme, low-priority flows can still maintain 90 percent of their average throughput of what it was without any flow control. This is not due to the fact that low-priority flows were not throttled enough to protect the flow rate of the high-priority flow. But, optimum controlling of the low-priority packet injection rate leads to less congestion and proper utilization of the medium, leading to negligible packet loss. This scheme maximizes the low-priority throughput also, and packets are injected at an optimized rate, which the network can handle at that point of time.

Under Mobility

The proposed protocol has been tested and evaluated under mobility of 0–10 mps to show the robustness of the proposed flow-rate control scheme even in continuously changing topology. Better performance can be achieved due to the fast back-propagation of congestion information to the low-priority source, which can adaptively take control decision of its flow. Figure 6.20(a) shows the performance of high-priority traffic in our proposed flow-rate control scheme in comparison with when (1) no LPCs are assigned to the low-priority flows and (2) the high-priority flow operates alone in the absence of any other flow. The high-priority flow operating alone yields a throughput of nearly

172 ■ Enhancing the Performance of Ad Hoc Wireless Networks

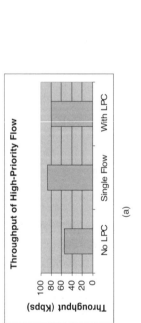
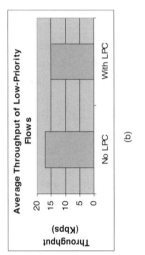

Figure 6.19 Comparison of throughput of (a) high-priority and (b) low-priority flows in different static scenarios.

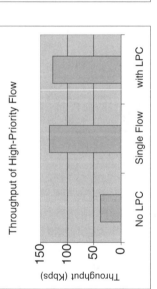
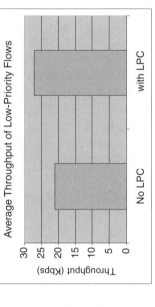

Figure 6.20 Comparison of throughput of (a) high-priority and (b) low-priority flows in different scenarios under mobility.

133 Kbps. When five other flows are introduced without LPC, the performance of this high-priority flow degrades to less than one-third of its previous value, as evident from Figure 6.20(a). With the proposed scheme of flow-rate control, throughput of this high-priority flow is incremented to nearly 127 Kbps.

Figure 6.20(b) shows the performances of low-priority flows in two scenarios: without any priority scheme (no LPC) and after introducing the proposed packet injection rate control (with LPC). The most interesting part of the evaluation shows that after introduction of the proposed scheme, low-priority flows can even improve their average throughputs as well. This improvement is possible due to optimal control of packet injection rate of low-priority flows, thus reducing congestion of the network, leading to optimal utilization of the medium with minimum packet loss.

6.5 Service Differentiation in Multi-Hop Intervehicular Communication

Intervehicular communication (IVC) on highways is one of the major application areas of ad hoc networks that enable multi-hop data exchange and forwarding between cars and between cars and stationary gateways. In an IVC scenario, some emergency situations on highways may require an immediate communication with police, hospitals, and highway assistance booth or with other cars. These messages should be forwarded on a top-priority basis to the destination for immediate attention. So it is evident that, in an IVC scenario, some message flows are to be treated as high-priority messages in order to ensure timely and reliable delivery. In the subsequent sections, we will discuss some of the routing issues in the IVC scenario and will use the notions of zone reservation and call blocking discussed above to propose a reactive routing scheme that will help us achieve priority-based service differentiation in a multi-hop IVC scenario.

6.5.1 Routing in an Unbounded Network

In this section, we discuss the major issues related to routing in an application like IVC, which is essentially an unbounded network. Subsequently, we illustrate the proposed reactive routing protocol in this context.

6.5.1.1 Bounded versus Unbounded Networks

Immense research has been done and several schemes have been proposed in the context of routing in wireless environments. But those schemes generally assume that the networks under consideration are bounded wireless networks. So, a required destination can be found out either proactively or reactively from that bounded region. In a proactive scheme each node in the network maintains the approximate network topology information through periodic exchange of some kind of control packets. To route a packet to a specific destination, each node just consults its topology information. On the other hand, reactive routing schemes search for a destination on demand basis through propagation of route request/route reply packets.

In an IVC scenario, it is not always possible to identify a destination in advance. For example, if an immediate communication with doctors is required on highways in case of any medical emergency, passengers of each car have to be interrogated to find whether a doctor is available in that car. If a doctor were found in any car, then that car would be the desired destination of the message. So, in IVC scenarios, content-based routing schemes[16] are generally employed, where a source node does not have any idea of the destination of the message beforehand. The nodes whose interests will be satisfied by the content of the message will be chosen as the destination. If the content satisfies the interests of more than one node, then there may be more than one destination of such messages. So, in such cases, a source node will issue a route request to search for a suitable destination of a message to its neighbors and each neighbor in turn forwards the same to their neighbors and so on. This way the route request gets forwarded in search of a destination. Each node on receiving the message will check whether the content of the message matches its interest. If that matches, then that node will be the prospective destination. It is evident from the above discussions that proactive routing protocols are difficult to apply in an IVC scenario because of the lack of prior knowledge about destination and the unbounded area of the network. The only possibility that can be explored in such scenarios is to reactively find out the suitable destination to route a packet.

So, here we propose a priority-based reactive routing technique for achieving service differentiation in multi-hop intervehicular communication.

6.5.1.2 Application-Dependent Route Discovery Process

In some applications like "looking for a doctor in nearby car for on-the-fly interactive consultancy" or "exchange of emergency information between police patrol cars on highways," the route request for the message should be transmitted omni-directionally by the sender issuing the emergency message. The omni-directional transmission of route request in such cases will enable the source node to search for the required destination (doctor/police) across the highway in both forward and backward directions wherever the destination is available.

However, information about a roadblock or accident ahead is required to be passed in the backward direction only to inform other vehicles along that route so that they can take an alternate route accordingly. Besides that, the nearby police station on the way back from the accident spot should be informed urgently for necessary assistance. These are the applications where backward transmission of route request is sufficient to find destinations.

On the contrary, if some information about a critical patient moving in a car is to be communicated with a nearby hospital for emergency hospitalization of the patient, then it is sufficient to transmit such route requests in a forward direction. In such cases, required destinations will be available in a forward direction.

So, the route-request forwarding techniques entirely depend on the nature of applications. The source node, requiring the communication, has to decide whether a particular application requires an omni-directional transmission of route requests or directional transmission in either a forward or backward direction. If an omni-directional transmission of route request is required for any application, then the source node only will broadcast the route request omni-directionally to all its neighbors. It is sufficient for other nodes, receiving the route request, to forward it in the reverse direction with respect to the direction of arrival of the request packet in order to maintain loop-free and possibly minimal forwarding paths for messages. Because nodes are equipped with directional antennas, forward or backward propagation in a single direction is possible, which minimizes contention in the medium.

In the proposed priority-based routing scheme, the selection of propagation process (forward/backward/omni) to be used for a route request issued by any source is only dependent on the requirement of the specific application. So, both high- and low-priority sources may use any of the above-mentioned propagation processes as necessary.

6.5.2 Implementation of Prioritized Routing Scheme in IVC Scenario

To implement the proposed priority-based routing scheme, each node periodically transmits an omni-directional beacon containing its node-ID and activity status, which can be either high or low. The default activity status of each node is low, which indicates that it is not within any high-priority zone. Each packet of a flow is tagged with the priority (high or low) of the corresponding flow. On receiving or overhearing a high-priority data packet, each node within the directional transmission zone of the sender toward the receiver sets its activity status as high. This status is retained for a predefined time interval, after which the node resets its status back to low to indicate that the node is no longer within a high-priority zone. Each node, on receiving a beacon from each of its neighbors, forms a table, known as the neighborhood direction status table (NDST). The NDST is derived from the angle signal table (AST) that is required to implement directional MAC protocol[14] and essentially contains the node-ID of the sender node, the direction of arrival of the beacon from the sender node, and the activity status of the sender.

According to our scheme, a source will initially send a route-request packet containing the subject of the message to search for suitable destinations. Whenever route-reply packets are propagated back by prospective destinations to the source issuing the route request, the source will select a suitable route. A source generally selects a shortest-cost route toward the chosen destination from the list of routes that are piggybacked with route-reply packets.

6.5.2.1 Route Computation and Zone Reservation by High-Priority Flows

In the case of high-priority messages, route-request packets are transmitted omni-directionally or directionally either in a forward or backward direction depending on the content of the message. Other nodes, receiving the request, will forward it in the reverse direction with respect to the direction of arrival of the request packet, thus minimizing the search space. If an intended destination is found on the way, then the destination will inform the high-priority source about the route to be taken to reach that destination. Among the several alternative routes available to the source to reach a destination, the shortest one will be selected.

As a high-priority flow is initiated, the nodes in the high-priority zone set their activity status as high. This information is eventually communicated to their neighbors through a beacon, which in turn updates their NDSTs. Thus each neighboring node becomes aware of the high-priority ongoing communication in the vicinity.

6.5.2.2 Route Computation and Adaptive Call Blocking by Low-Priority Flows

Any node handling a low-priority flow selects a node m as its next hop toward the intended destination, only if the directional transmission zone from that node to m does not contain any node with high activity status in its NDST. To select such a route avoiding a high-priority reserved zone, a low-priority source will essentially transmit its route-request packet, using a directional antenna, in all the sectors excluding the sectors containing active nodes. It is possible to form such multiple beams along different sectors using steerable beamforming antennas. A node, on getting the low-priority route request, will forward the request packet as before in all sectors except the sectors containing high-priority nodes. Thus, each route-reply packet, sent by a destination to the low-priority source, would contain a route that avoids the high-priority zone.

If no such route is available for routing low-priority traffic, then the source node will temporarily block the low-priority flow until the high-priority zone is released by high-priority flow. The low-priority source will check the NDST continuously to resume its transmission. Whenever a high-priority communication within a high-priority zone is absent for a considerable period of time, then the nodes belonging to that zone set their activity statuses as low and a blocked low-priority source may reinitiate its route discovery process through that zone.

If an intermediate node on a low-priority route senses that a new high-priority communication has been initiated in its vicinity and the path to be taken by the node toward destination has to pass through the newly formed high-priority zone, then the node will try to rediscover a route avoiding that zone. If it fails to find such a route it will send an error signal to inform the source to discover a new route toward destination. The low-priority source node tries to find a new route avoiding the high-priority zone. If it is not found, the source will temporarily block the flow. Figures 6.21(a) and 6.21(b) illustrate the advantage of using directional antennas in an IVC scenario to

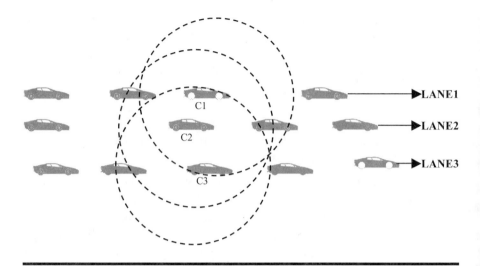

Figure 6.21(a) Overlapping transmission zones of cars C1, C2, and C3 using omni-directional antennas inhibit the possibility of simultaneous communications among the vehicles along lanes 1, 2, and 3.

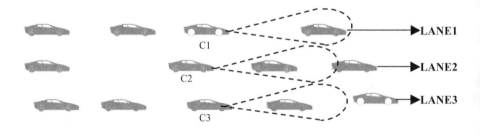

Figure 6.21(b) Non-overlapping transmission zones of cars C1, C2, and C3 using directional antennas enhance the possibility of simultaneous communications among the vehicles along lanes 1, 2, and 3.

enhance SDMA efficiency. Figures 6.21(c) and 6.21(d) illustrate, respectively, the scenarios when zone reservation and call blocking are essential.

6.6 Discussion

In this chapter, we suggested a zone-reservation-based mechanism toward prioritized routing with the objective of providing an interference-free communication to high-priority traffic. To improve the high-priority throughput further, we suggested adaptive call blocking of

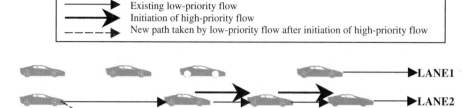

Figure 6.21(c) Initiation of a high-priority flow from a car in LANE2 handling an existing low-priority communication causes the low-priority flow to take a diverse route along LANE3.

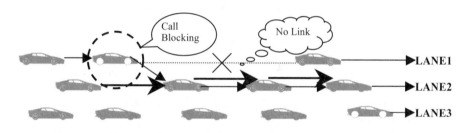

Figure 6.21(d) Initiation of a high-priority flow from a car in LANE2 handling an existing low-priority communication forces the low-priority flow to block itself due to the absence of any other alternative path.

low-priority flows. Unless narrow-beam directional antennas are used, it is not possible to accommodate multiple numbers of non-overlapping high-priority zones. However, the paths would become less stable with narrow-beam directional antennas, if the nodes are mobile.[17] Therefore, it is imperative to have an adaptive call-blocking mechanism to ensure good performance of high-priority flow under a heavy traffic scenario. We have used the notion of zone reservation by high-priority flows and call blocking by low-priority flows to devise a reactive routing strategy, which is suitable for the unbounded network scenario of IVC. Future investigations can focus on the impact of adaptive beamwidth and transmission power control of directional antennas to improve the throughput of prioritized flow without degrading the low-priority flow to a large extent.

In this context, we have also studied the flow-rate control by the detection of flows in the nodes where actual congestion is created by route coupling. So far, we have implemented the flow-rate control mechanism where the high-priority flow interacts with low-priority flows. This mechanism can be extended to cope with the scenario where high-priority flows contend among themselves for access to the medium. Also, by this mechanism, we are trying to improve fairness among the low-priority flows when they contend among themselves for access to the medium in absence of the high-priority flow.

References

1. Wu, K. and Harms, J., QoS Support in Mobile Ad Hoc Networks, in *Crossing Boundaries—An Interdisciplinary Journal*, Vol. 1, No. 1, Fall 2001.
2. Chakrabarti, S. and Mishra, A., QoS Issues in Ad Hoc Wireless Networks, in *IEEE Communications Magazine*, Vol. 39, No. 2, February 2001, p. 142.
3. Demetrios, Z.Y., A Glance at Quality of Service in Mobile Ad Hoc Networks, http://www.cs.ucr.edu/~csyiazti/cs260.html. Final research report for cs260—Seminar on Mobile Ad Hoc Networks, Fall 2001.
4. Chen, S. and Nahrstedt, K., Distributed Quality-of-Service Routing in Ad Hoc Networks, in *IEEE Journal on Selected Areas in Communications*, Vol. 17, No. 8, August 1999, p. 1488.
5. Xiao, H. et al., Flexible Quality of Service Model for Mobile Ad-Hoc Networks, in *Proc. IEEE VTC 2000-spring*, Tokyo, Japan, May 2000.
6. Mohapatra, P., Li, J., and Gui, C., QoS in Mobile Ad Hoc Networks, Special Issue on QoS in Next-Generation Wireless Multimedia Communications Systems, in *IEEE Wireless Communications*, Vol. 10, No. 3, p. 44.
7. Pallot, X. and Miller L.E., Implementing Message Priority Policies over an 802.11 Based Mobile Ad Hoc Network, in *Proc. MILCOM 2001*, Washington, D.C., October 2001.
8. Wu, K. and Harms, J., On-demand Multipath Routing for Mobile Ad Hoc Networks, in *Proc. EPMCC 2001*, Vienna, February 20–22, 2001.
9. Pearlman, M.R. et al., On the Impact of Alternate Path Routing for Load Balancing in Mobile Ad Hoc Networks, in *Proc. MobiHOC 2000*.
10. Bandyopadhyay, S. et al., Multipath Routing in Ad Hoc Networks with Directional Antenna, in *Proc. IFIP TC6/WG6.8 Conference on Personal Wireless Communications (PWC 2002)*, Singapore, October 2002.
11. http://www.et2.tu-harburg.de/fleetnet/
12. Nandagopal, T. et al., Service Differentiation through End-to-end Rate Control in Low Bandwidth Wireless Packet Networks, in *Proc. 6th International Workshop on Mobile Multimedia Communications*, San Diego, California, November 1999.

13. Kang, S.S. and Mutka, M.W., Provisioning Service Differentiation in Ad Hoc Networks by the Modification of Backoff Algorithm, in *Proc. International Conference on Computer Communication and Network (ICCCN) 2001*, Scottsdale, Arizona, October 2001.
14. Ueda, T. et al., A Rotational Sector-based, Receiver-oriented mechanism for Location Tracking and Medium Access Control in Ad Hoc Networks Using Directional Antenna, in *Proc. IFIP conference on Personal Wireless Communications PWC 2003*, Venice, Italy, September 23–25, 2003.
15. Roy, S. et al., A Network-aware MAC and Routing Protocol for Effective Load Balancing in Ad Hoc Wireless Networks with Directional Antenna, in *Proc. ACM MobiHoc, 2003*, Maryland, June 1–3, 2003.
16. Carzaniga, A., Rosenblum, D.S., and Wolf, A.L., Content-based Addressing and Routing: A General Model and Its Application, Department of Computer Science technical report CU-CS-902-00, University of Colorado, January 2000.
17. Bandyopadhyay, S. et al., Improving System Performance of Ad Hoc Wireless Network with Directional Antenna, in *Proc. IEEE International Conference on Communications (ICC 2003)*, Anchorage, Alaska, May 11–15, 2003.
18. Kolarov, A. and Ramamurthy, G., A Control Theoretic Approach to the Design of Close Loop Rate Based Flow Control for High Speed ATM Networks, in *Proc. IEEE INFOCOM'97*, April 1997, p. 293.
19. Keshav, S., A Control-theoretic Approach to Flow Control, in *Proc. ACM SIGCOMM'91*, Zurich, Switzerland, September 1991, p. 3.
20. Kalmanek, C. R., Kanakia, H., and Keshav, S., Rate Controlled Servers for Very High Speed Networks, in *Proc. Globecom 1990*, December 1990.
21. Benmohamed, L. and Meerkov, S.M., Feedback Control of Congestion in Packet Switching Networks: The Case of a Single Congested Node, in *IEEE/ACM Trans. of Networking*, Vol. 1, No. 6, 1993, p. 693.
22. Kang, S.-S. and Mutka, M. W., Provisioning Service Differentiation in Ad Hoc Networks by the Modification of Backoff Algorithm, in *Proc. International Conference on Computer Communication and Network (ICCCN) 2001*, Scottsdale, Arizona, October 2001.
23. Mangold, S. et al., IEEE 802.11e Wireless LAN for Quality of Service (invited paper), in *Proc. European Wireless 2002*, Florence, Italy.
24. Yang, X. and Vaidya, N.H., Priority Scheduling in Wireless Ad Hoc Networks, in *Proc. ACM International Symposium on Mobile Ad Hoc Networking and Computing (MobiHoc)*, June 2002.
25. VanDoren, V.J., Understanding PID Control: Familiar Examples Show How and Why Proportional-Integral-Derivative Controllers Behave the Way They Do, *Control Engineering* online, www.controleng.com, June 1, 2000.

26. Minorsky, N., Directional Stability of Automatically Steered Bodies, in *J AM Soc Naval Eng*, Vol. 34, 1922, p. 284.
27. Franklin, G.F., Powell, J.D., and Emami-Naeini, A., *Feedback Control of Dynamic Systems*, Addison-Wesley, Singapore, 1988.
28. Ueda, T. et al., Evaluating the Performance of Wireless Ad Hoc Network Testbed with Smart Antenna, in *Proc. Fourth IEEE Conference on Mobile and Wireless Communication Networks (MWCN2002)*, September 2002.
29. QualNet Simulator Version 3.1, Scalable Network Technologies, www.scalable-networks.com

Chapter 7
Conclusion

MANET routing protocols have been extensively studied in the past by several researchers. However, it has been observed that none of the existing routing protocols have explored the effect of mutual interference between ongoing communications during routing, which is a key factor for the degradation of overall network performance. Moreover, most of the existing routing protocols in wireless ad hoc networks use omni-directional antennas and the wide transmission zone of omni-directional antennas also increases the possibility of mutual interference between communicating nodes. So, the performance gets even more affected. So, coupling between routes within the proximity of each other in a radio environment has motivated the authors to explore the idea of zone-disjointness.

It is evident from the discussions in the previous chapters that the use of directional antennas in the context of ad hoc wireless networks can largely reduce radio interference, thereby improving the utilization of wireless media. Several directional MAC protocols have been proposed to exploit this advantage and improve SDMA efficiency. However, even if we have an efficient directional MAC protocol, it alone would not be able to guarantee good system performance, because coupling between two interfering routes (which are apparently disjoint but physically located close enough to interfere with each other) may not allow the two routes to communicate simultaneously even if directional antennas are employed. Selection of routes avoiding mutual interference among selected routes is very important in this context.

Thus, it is evident that we have to have a proper routing strategy in place that exploits the advantages of directional antennas in order to improve overall network performance using directional antennas.

However, finding the desired direction of transmission or reception for communicating using directional antennas is a complex issue. So, in this work, we have initially devised a location-tracking mechanism using directional antennas. Subsequently, a receiver-centric MAC protocol has been proposed, where each node keeps certain neighborhood information dynamically so that each node can avoid interference by keeping track of other communicating nodes at that instant of time. On top of that an adaptive routing strategy is proposed that exploits the advantages of directional antennas in ad hoc networks through the selection of maximally zone-disjoint shortest routes. Zone-disjoint routes would minimize the effect of route coupling in wireless media by selecting routes in such a manner that data communication over one route will minimally interfere with data communication over the others. The proposed routing strategy is proactive and network-aware. Each node in the network maintains the current network status information (i.e., approximate global topology information and the information of the nodes currently involved in some communications). This information helps a node to find a route while minimally disturbing other ongoing communications in the vicinity and that in turn ensures effective load balancing in the wireless medium. This scheme is applied to design and implement both single-path and multipath routing protocols on ad hoc networks with directional antennas. The routing schemes for ad hoc networks usually employ single-path routing, which might not ensure desired end-to-end delay. However, once a set of paths between source s and destination d is discovered, in some cases it is possible to improve end-to-end delay by splitting the total volume of data into separate blocks and sending them via selected multiple paths from s to d, which would eventually reduce congestion and end-to-end delay. The most significant challenge to making the use of multipath routing protocols effective in wireless environments involves considering the effects of route coupling. So, we have applied the notion of zone-disjoint routing to design a multipath routing protocol that in turn provides better load balancing and reduces end-to-end delay.

Quality-of-service (QoS) provisioning is a new but challenging research area in the field of mobile ad hoc networks to support multimedia data communication. However, the existing QoS routing protocols in ad hoc networks did not consider the effect of route coupling that can be avoided by using zone-disjoint routes. In this work, we have also discussed a scheme for supporting priority-based QoS in MANET by classifying the traffic flows in the network into different priority classes and giving different treatment to the flows belonging to different classes during routing so that the high-priority flows will achieve the best possible throughput. The underlying objective is to reduce the effect of coupling between routes used by high- and low-priority traffic by reserving the zone of communication. The part of the network, used for high-priority data communication, that is, the high-priority zone, will be avoided by low-priority data through the selection of a different route that is maximally zone-disjoint with respect to high-priority zones and that consequently allows contention-free transmission of high-priority traffic. The suggested protocol selects the shortest path for high-priority traffic and diverse routes for low-priority traffic that will minimally interfere with high-priority flows, thus reducing the effect of coupling between high- and low-priority routes. We have further investigated the QoS provisioning issues by adaptively maximizing low-priority flows while maintaining high-priority flows at a desired level so that full utilization of wireless media is achieved through adaptive rate control. To provide this desired service differentiation to high-priority flows, we need a flow-control algorithm, where the low-priority flows, causing interference to a high-priority flow, detect and measure high-priority flow rate at each node on their routes and consequently adjust their flow rates using a control-theoretic approach to protect the high-priority flow at its desired level. This detection and measurement is done at the MAC layer of each node participating in communication.

Intervehicular communication (IVC) on highways is one of the major application areas of ad hoc networks that enable multi-hop data exchange and forwarding between cars and between car and stationary gateways. In an IVC scenario, some emergency situations on highways may require an immediate communication with police, hospitals, a highway assistance booth, or with other cars. These messages should be forwarded on a top priority basis to the destination for immediate attention. So it is evident that, in an IVC scenario, some message flows are to be treated as high-priority messages to ensure timely and reliable delivery. This application scenario has motivated us to apply the concept of a priority-based zone-reservation scheme in the IVC scenario and to formulate a reactive routing scheme suitable for this scenario.

In this book, we have also shared our experience during the realization of an ad hoc network testbed using a directional ESPAR antenna where we have implemented a location-tracking mechanism that does not require any common clock synchronization and a directional MAC as well as a congestion-aware zone-disjoint routing protocol.

Transmission beamwidth (θ) and transmission range (R) are two important parameters that need to be optimized to restrict congestion and to handle mobility in the context of directional communication. A smaller value of θ will definitely improve SDMA efficiency as well as bridge the gap in a sparse network with its longer range but at the same time under mobility there will be a high chance that a mobile node will move out of the narrow transmission beam, causing frequent route error. The longer transmission range has another disadvantage; it may often create interference to some faraway communication. Therefore, adaptive control of θ and R may help to reduce congestion, ensure network connectivity, and, consequently, improve routing performance. Moreover, adaptive transmission range will ensure balanced consumption of battery power of the power-limited mobile hosts. Additionally, the narrow beamwidth of directional antennas may also improve multipath routing efficiency under high node density and will allow more high-priority flows in a system in the context of priority-based QoS. Power-aware protocols in this context require further investigations.

Index

A

ABF, 60
   architecture, 62
ABR, 30
Acceptable power to send. *See* APTS
ACK reception, PCMA and, 25
ACR routing protocols
   evaluation under mobile scenarios, 130
   evaluation under static scenarios, 128
Active Bat Location System, 81
Active Campus, 81
Active directional neighbors, 106
Active node list. *See* ANL
AD converters, 60
Ad hoc networks, 1
   applications of, 2
   characteristics of, 2
   evaluation of protocol performances in, 41
   hybrid, 4
   limitations and properties of, 27
   location estimation in, 84
   multipath routing protocols for, 119
   performance of, 8
   prospective usages of, 3
   quality of service in, 136
   routing in, 6, 27

   smart antennas in, 64
   use of directional antennas in, 76
Ad Hoc On-demand Distance Vector. *See* AODV
Ad hoc Protocol Evaluation testbed. *See* APE testbed
Adaptive array antennas, 5, 55
   advantages of, 64
   designing MAC protocols using, 15
Adaptive beam controls, 58, 65, 179
Adaptive beam patterns, 57
Adaptive call blocking, 141
   low-priority flows, 177
Adaptive Communications Research Laboratories. *See* ATR-ACR
Adaptive flow rate control, 166
Adaptive Location-Aided Routing from Mines. *See* ALARM
Adaptive route calculation, 140
Adaptive routing strategy, 101, 124, 129, 132
   route selection, 118
Adaptive transmission range, 186
Aerial beamforming. *See* ABF
Affinity, 33
Aged packets, 42
ALARM, 35
ALOHA, 16

Alternate Path Routing. *See* APR
Analog-digital converters, 60
Analytical modeling, 8
Angle signal table. *See* AST
Angle spread, 64
Angle-of-arrival measurements. *See* AoA measurements
Angulation, 84
ANL, 106, 109, 123, 126
  formation of, 111
Antennas. *See also* specific antenna types
  alignment of, 85
  synchronization by nonprimary reference nodes, 87
AoA measurements, 80, 85
AoA-based location estimation technique, 82
AODV, 27, 29, 42, 105, 128, 129
  communication gray zones in implementations of, 8
  testbed testing of, 46
APE testbed, 47
APE-view, 47
Application layer metrics, 44
Application-dependent route discovery process, 175
APR, 119
APTS, 25
Associativity ticks, 30
Associativity-Based Routing. *See* ABR
AST, 23, 57, 109, 176
  formation of, 83, 110
ATR-ACR, 56, 64
Attenuation, distance-based, 41
Average throughput, 72

B

Back-off schemes, 44, 137
BAT system, 81
Beacons, 30, 66, 80
  link-level, 31
Beam controls, capacity increase through smart antennas, 58
Beamforming
  directional, 55
  neighbor discovery with, 48
Blind neighbor discovery, 48
Bounded networks, routing in, 174

Breadcrumb routers, 4
BTMA, 16
Buffer zone effect, 31
Busy Tone Multiple Access. *See* BTMA

C

Caching, 28
Call blocking, 149
  adaptive, 141
  effectiveness of, 145
Carrier Sense Multiple Access. *See* CSMA
Carrier sense multiple access/collision avoidance. *See* CSMA/CA
Carrier sense multiple access/collision detection. *See* CSMA/CD
Carrier sensing, 18, 44
Channel distortion, 58
Channel vulnerability, security issues and, 8
Clear to send. *See* CTS
Click, 47
CMA, 64
Co-channel interference, 57, 64
Collision avoidance, 13
Common nodes, 114
Communication paths, 107
Community networks, 4, 80
Computational limitations, security issues and, 8
Congestion-related metrics, 42
Constant modulus algorithm. *See* CMA
Content-based routing schemes, 174
Contention window, 137, 153
Control overhead, 126
Control-theoretic approach, 150, 152, 154
Controlled MAC, 6
Coordinate calculation process, 86
Coordinate synchronization, 82
Correctness, degree of, 111
Correlation factor of a route, 108, 115
Coupled flow, 147
Cricket location support system, 81
CSMA, 16
CSMA/CA, 18
CSMA/CD, 18
CTS, 6, 66, 152
  directional (*See* DCTS)

Index ■ 189

D

Data communication, 57
DBF, 59
  architecture, 61
DBTMA, 16, 18, 44
  directional version of, 24
DBTMA/DA, 44
DCF, 17
DCTS, 19
DDSR, 40
Deafness, 22, 40
Decentralized operation, 2
Delay, 42
  spread, 58, 64
Destination Sequenced Distance Vector.
  See DSDV
Detected maximum packet arrival interval.
  See DMPAI
Digital beamforming. See DBF
DiMAC, 40
Direction of arrival estimation algorithm.
  See DoA estimation algorithm
Directional antennas, 2, 55, 132, 183
  adaptive routing strategy for, 101
  advantages over omni-directional
    antennas, 45
  collision avoidance schemes using, 15
  estimating node location using, 79
  link characterization, 48
  MAC protocols with, 19, 44
  multipath routing using, 119
  narrow-beam, 179
  null steering of, 23
  patterns of, 75
  performance of, 63
  power-control schemes using, 25
  routing protocols using, 38, 114
  use in ad hoc wireless networks, 76
  use of maximally zone-disjoint paths
    with, 130
  use of priority-based routing with, 137
  zone-disjoint routing and, 104
Directional beam patterns, 57
Directional beamforming, 55
Directional CTS. See DCTS
Directional media access control
  protocols, 66
Directional NAV, 24

Directional network allocation vector. See
  DNAV
Directional routing protocols, 79
Directional RTS. See DRTS
Disaster relief operations, use of ad hoc
  networks in, 4
Distance Routing Effect Algorithm for
  Mobility. See DREAM
Distance vector routing, 127
Distance-based attenuation, 41
Distance-based triangulation technique,
  82
Distributed coordination function. See
  DCF
Distributed infrastructure-free positioning
  algorithm, 81
DMPAI, 162
DNAV, 67
DoA estimation algorithm, 82
Doppler spread, 64
Down converters, 60
DREAM, 35
DRTS, 19
DRTS/DCTS-based DMAC, 21
DRTS/oCTS-based DMAC, 21
DSDV, 27, 35, 43
  testbed testing of, 46
DSR, 27, 42, 45, 119
  on-demand directional, 114
Dual Busy Tone Multiple Access. See
  DBTMA
Dynamic Source Routing. See DSR
Dynamic topology changes, 27

E

E-MAC, 73
  performance evaluation of, 75
EIFS, 25
Electronically Steerable Passive Array
  Radiator antenna. See ESPAR
  antenna
Emergency response networks, 4, 136
End-to-end delay performance, 43, 72,
  137, 184
  improvement with multipath routing,
    119
End-to-end flow control, 151
Energy saving, use of PCMA for, 25

Enterprise networks, 4
Environmental sensing network, 4
ESPAR antenna, 23, 56, 59, 124, 164, 185
   adaptive routing with, 129
   configuration of, 62
   multipath routing with, 129
   performance evaluation, 71
   priority-based QoS routing simulation, 143
   shortest-path routing with, 128
   synchronization using nonprimary reference node, 87
Expanding ring search, 29
Exposed terminals, 6, 14, 44

F

Fading models, 57
Fairness, 150
FAMA, 16, 19, 44
Fast fades Doppler spread, 64
Feedback control, 150, 154
   low-priority flow rate, 163
Fisheye State Routing. *See* FSR
Fixed-beam antennas, 5, 55, 59
Flooding, 29, 114, 127
Floor Acquisition Multiple Access. *See* FAMA
Flow-rate control, 155, 179
   priority-based, 150
Forwarding, 48
Frequency converters, 60
Frequency reuse, 57
Frequency utilization, 57
FSR, 37, 128

G

Gedir protocol, 35
Geographic Distance routing protocol. *See* Gedir protocol
Geographical Routing Algorithm. *See* GRA
Global link state table. *See* GLST
Global positioning system. *See* GPS
Global route search, 29
GloMoSim, 9
GLST, 110, 124
   formation of, 112
GPS, 34, 48, 65, 79

GPSR algorithm, 35
GRA, 35
Greedy Perimeter Stateless Routing algorithm. *See* GPSR algorithm
GRID protocol, 35

H

Hazy State Routing Protocol. *See* HSLS
Hello messages, 36, 66
Heterogeneity of ad hoc nodes, 3
Hidden terminals, 6, 14, 22, 44, 70
   problems of CSMA, 16
High-priority flow, 135, 140, 148, 185
   detecting and measuring rates of, 159
High-priority zone, definition of, 138
Hop count, 30, 46, 114
HSLS, 48
Hybrid ad hoc networks, 4

I

IEEE 802.11, 8, 17, 40, 44, 57, 66, 71, 128, 164
IFS, 153
Indoor localization, using RADAR system, 81
Informed neighbor discovery, 48
Infrared transmissions, 17
Infrastructure, security issues due to absence of, 8
Infrastructure-free positioning algorithm, distributed, 81
Intelligent digital signal processing algorithms, 58
Inter frame spaces. *See* IFS
Inter symbol interference. *See* ISI
Interference, 40, 41, 64
   co-channel, 57
   mutual, 135
Intervehicular communication. *See* IVC
Intruding terminal problem, 19
ISI, 58
IVC, 136, 185
   service differentiation in multi-hop scenario, 173

L

LAR, 34, 79

Latency, 47
Lateration, 80, 84
Link asymmetry, 3
Link characterization, 48
Link quality, 58
Link stability, 33
Link weight, 141
Link-cost, 115
Link-state reduction, 36
Link-state routing, 127
Link-state table, 37
Load balancing, 101, 114, 184
　improvement with multipath routing, 119
　zone-disjoint routes and, 103
Location estimation, 79, 84
　error in, 98
　multi-hop, 90
　multi-hop with secondary reference, 95
　single hop with two reference nodes, 93
　using reference nodes, 86
Location systems, 81
Location tracking, 22, 82, 170, 184
　evaluating, 91
　mechanisms, 56
　　neighborhood discovery, 65
Location-Aided Routing. *See* LAR
Location-discovery techniques, 80
Location-estimation protocols, 82
Loss rate, 45, 47
Low-priority flow, 135, 148
　feedback control of rate of, 163
Low-priority-flow-controller. *See* LPC
LPC
　evaluation of, 171
　performance of, 165, 168
Lucent WaveLAN-II radios, 46

M

MAC, 137, 170
　performance comparisons, 44
　power-controlled, 24
　protocols, 2, 183
　　adaptive, 23
　　designing, 13
　　directional (*See* DiMAC)
　　on-demand schemes, 23
　　scheduled schemes, 23
　　using directional antennas, 19, 67
　QoS in, 153
　types of, 6
MAC-based location-tracking
　　mechanisms, 64
MAC/DA1, 44
MAC/DA2, 44
MAC/DA2ACK, 44
MACA, 16, 19, 44
MACAW, 16, 19, 44
Macfilter, 46, 95
MANET
　emulation using Macfilter, 46
　QoS in, 153
　routing protocols, 183 (*See also* Routing)
　unicast routing protocols, 35
Mapping tables, 89. *See also* specific tables
Max-min flow control, 157
Maximally zone-disjoint multipath
　　routing, 119
　selection of, 123
Maximally zone-disjoint shortest path
　　routing, 114, 184
Maximally zone-disjointness, 142
Maximum cross correlation coefficient.
　　See MCCC
MCCC, 64
Media Access Control. *See* MAC
Media Access Protocol for Wireless LANs.
　　See MACAW
Medium utilization, improving, 15
Microcell, 57
Military sensor networks, 4
Minimum mean square error criterion. *See* MMSE
MMAC, 22
MMSE, 64
Mobility, 2
Mobility rate, 43
Modeling, 8
Movement speed, 43
MPR, flooding mechanism, 36
Multi-hop routing, 2, 44, 58
　efficiency of, 71
　service differentiation in IVC, 173
　variation in link quality during, 45
Multi-hop RTC MAC. *See* MMAC

Multi-hop-based location computing, 82
Multipath fading, 41
Multipath propagation, improving link quality in, 58
Multipath reflections, 24
Multipath routing protocols, 43, 103, 129, 132, 184
    load balancing through, 123
    performance evaluation, 124
    priority-based, 132
    using omni-directional and directional antennas, 119
Multiple Access with Collision Avoidance. *See* MACA
Multipoint relay. *See* MPR
MUSIC location-tracking mechanisms, 64
Mutual interference, 135, 137

N

N-BF, 48
NANL, 109
    formation of, 110
NAV, 24
NDST, 176
Neighbor-sensing, 36
    beamforming and, 48
Neighborhood active node list. *See* NANL
Neighborhood direction status table. *See* NDST
Neighborhood discovery, 82
    location-tracking mechanisms for, 65
Neighborhood link state table. *See* NLST
Neighborhood tracking, 79
Neighbors
    categories of, 48
    definition of, 70
Network allocation vector. *See* NAV
Network awareness, 104, 109, 184
Network connectivity, 45
Network infrastructure, viability of, 3
Network topology, dynamically changing, 8
Network-aware routing, 170
    maximally zone-disjoint shortest path, 113
NLST, formation of, 82
Nodes
    affinity, 33

communication between, 56
estimating location using directional antennas, 79
heterogeneity in ad hoc networks, 3
in ad hoc networks, 1
intermediate, routing decisions, 118
mobility, 25
mobility rate, 43
primary reference, 85
secondary reference, 85
timer-based states, 29
topology-aware, 109
vulnerability of, 8
Nonprimary reference nodes, 87
NS-2, 9, 41
Null controls, capacity increase through smart antennas, 58
Null steering, 65

O

OLSR, 36
Omni-directional antennas, 2, 55, 68. *See also* ESPAR antenna
    advantages of directional antennas over, 45
    collision avoidance schemes using, 15
    load balancing using, 103
    MAC protocols with, 15, 44
    multipath routing using, 119
    power-control schemes using, 24
    routing protocols using, 27
    use of zone-disjoint paths with, 137
Omni-directional RTS. *See* oRTS
On-demand reactive routing schemes, 126
On-demand routing, 38, 43
    analysis of, 42
One-hop average end-to-end delay, 72
One-hop neighbors, 36
One-hop traffic, 44
OPNET, 9
Optimized Link State Routing. *See* OLSR
oRTS, 19
oRTS/oCTS-based DMAC, 21

P

Packet arrival interval. *See* PAI
Packet arrival interval table. *See* PAIT

Packet delivery ratio, 153
Packet forwarding technique, 57
Packet injection interval. See PII
Packet injection rate. See PIR
Packet-level simulation model, 43
PAI, 161
PAIT, 161
Parasitic elements, 60
Passive radiators, 60
Path length, 104
Path loss, 24
Path stability, 32
PCMA, 24
Performance evaluation techniques, 8, 143, 164
　　ESPAR antennas, 71
Periodic table exchange, 37
PHY layer of wireless networks, 17
Physical carrier sensing, 44
PID control, 154, 157
PII, 163
　　control of, 165
PIR, 163
PM Table, formation of, 89
Post-Synchronization Mapping Table. See PM Table
Power conservation, 7
Power control, 65
Power Controlled Multiple Access protocol. See PCMA
Power limitations, security issues and, 8
Power-aware routing, 6
Power-control mechanisms, 24
　　two-level transmit, 26
Powerwave, 46, 95
PPAI, 163
Predecessor nodes, 29
Primary reference nodes, 85
Priority-based flow control strategies
　　use of directional antennas with, 159
　　use of PID controller with, 155
Priority-based routing, 135
　　implementation in IVC scenario, 176
　　multipath, 132
　　reactive, 174
Proactive routing protocols, 6, 35, 48, 127, 184
　　performance evaluation of, 43

Propagated packet arrival interval. See PPAI
Propagation
　　degradation of wireless signal during, 24
　　process selection, 175
　　RREQ, 39
Proportional-integral-derivative control. See PID control

Q

QoS, 135, 184
　　priority-based flow-rate control for, 150
　　routing, 7, 137
　　　　priority-based using zone reservation, 139
　　support schemes, 46
Quality of service. See QoS
QualNet, 9, 41, 44, 57, 63, 71, 75, 105, 124, 143, 164
Quasi-switched-beam antenna, use of ESPAR as, 60
Query dissemination, 127
Query-reply process, 31
Queue lengths, 42

R

RADAR system, 81
Radiators, 60
Radio frequency transmissions, 17
Radio interference, reduction with directional antennas, 2, 19
Radio propagation, 57
Random access channels, 26
Random MAC, 6
Rayleigh distribution, 57
Reactive routing protocols, 6, 27
　　directional, 40
　　on-demand, 126
　　performance evaluation of, 43
　　priority-based, 174
Received signal strength measurements. See RSSI measurements
Receiver-oriented location-tracking mechanism, 65
Receiver-Oriented Multiple Access. See ROMA

Recency of information, 111
Reception range, 45
Reference nodes, 85, 90
   location estimation by use of, 86
   nonprimary, 87
Request power to send. See RPTS
Request to send. See RTS
Request zone, 34
Reserved node list. See RNL
Resource reservation signaling, QoS, 136
Reuse distance, 57
Ricean distribution, 57
RNL, 139
ROMA, 22
Rotational-sector-based mechanisms, 67, 170
Route cache, 28, 39, 127
Route computation
   high-priority flows, 140, 176
   low-priority flows, 141, 177
Route coupling, 102, 106, 114, 119, 135, 165, 184
   load balancing with, 103
Route discovery, 15, 27, 30
   application-dependent process, 175
   latency, 40
Route failures, 128
Route maintenance, 27, 127
   minimization by selection of stable routes, 32
   preemptive, 47
Route reconstruction, 39
Route selection criteria, 101, 183
Routers, breadcrumb, 4
Routes, freshness of, 42
Routing, 6, 170
   ACR, evaluation of under static scenarios, 128
   AODV, 29
   distance vector, 127
   distribution for load balancing, 103
   link-state, 127
   load, 43
   location-aided, 79
   maximally zone-disjoint multipath, 119
   maximally zone-disjoint shortest-path, 114
   network-aware, 113
   on-demand reactive, 126

performance comparisons, 42
protocols, 2, 183 (See also specific protocols)
   content-based, 174
   directional, 79
   priority-based, 135, 139, 174 (See also Priority-based routing)
   proactive, 35
   reactive, 27
   table-driven, 104
   unicast, 35
   using directional antennas, 38
   using omni-directional antennas, 27
strategy, 114
table entries, 29
techniques for improvement using directional antennas in MANETs, 41
RPTS, 25
RREQ propagation, 39
RRT, 161
RSSI measurements, 80
RTS, 6, 66, 152
   directional (See DRTS)
   omni-directional (See oRTS)
RTS-CTS-DATA-ACK, 18, 66, 159
RTS-Reception-Time. See RRT
RTS/CTS, 65
   floor reservation scheme, 55

S

Scalability, 132
Scalable routing, 6
SDMA, 55, 129, 186
   by smart antennas, 58
Search-and-rescue operations, 4
Searches, 29
Secondary reference nodes, 85
Security, 8
Sensor networks, 4
   remote environment monitoring with, 80
Service differentiation, 136, 151
   in multi-hop intervehicular communication, 173
Shadowing, 24, 41
Shortest-path protocols, 43, 114, 128
Signal attenuation-based measurement, 81

Index

Signal Stability-Based Adaptive Routing. *See* SSAR
Signal to interference and noise ratio. *See* SINR
Simulation modeling, 8, 42
   multipath routing, 124
   packet-level, 43
Simultaneous communication, 44, 58
Single-path routing, 102, 104, 132, 184
   adaptive, 120
SINR, 5, 57
Slow fades delay spread, 64
Smart antennas, 57
   classification of, 59
   directional, 2 (*See also* Directional antennas)
   use in ad hoc networks, 5
   features of, 58
   issues in wireless ad hoc networks, 64
SMR, 119
Source routing, 29, 44
Space Division Multiple Access. *See* SDMA
Spatial channel utilization, improvement with DRTS, 21
Spatial reuse, 22, 44
Split Multipath Routing. *See* SMR
Split-channel Reservation Multiple Access. *See* SRMA
SpotON ad hoc location system, 81
Spread, 64
SRMA, 16
SSAR, 31
Stability-based routing, 32
Stable-path algorithm, 33
Stale packets, 43
Static routes, 44
Steady-state performance, 43
Steerable adaptive array antennas, 5, 59
Suboptimal routes, 41
Suitable next hop, 115
Surveillance networks, 4
Sweeping, 38, 40
Switched-beam antennas, 5, 59
   use of ESPAR as, 60

T

T-BF, 48

Table-driven adaptive routing, 104, 132. *See also* Routing
TDoA measurements, 80
Testbeds, 8, 41, 80
   APE, 47
   ESPAR antenna, 56, 64
   location tracking, 91
   MANET, 46
   performance evaluation using, 45
Throughput, 42, 44, 72, 137
Time-difference-of-arrival measurements. *See* TDoA measurements
Time-of-arrival measurements. *See* ToA measurements
ToA measurements, 80
Topology control, use of PCMA for, 25
Topology discovery, 36
Topology maintenance, 15
Topology maps, 37
TORA, 43
TR-BF, 48
Tracking mechanisms, 56
Traffic flows, 120
Traffic sensor networks, 4
Transmission beamwidth, 105, 138, 160, 186
Transmission power-control, 25, 58, 179
Transmission range, 2, 6, 8, 13, 15, 22, 24, 26, 33, 38, 45, 70, 82, 86, 105, 129, 138, 159, 186
   omni-directional, 120, 125
Transmission zone, 105, 160
Triangulation, 80, 82, 84
Two-hop neighbors, 36

U

UDAAN, 48
Unbounded networks, routing in, 174
Unicast routing protocols, 35
   analysis of, 42
Uppsala University APE testbed, 8

V

Variable capacity links, 3
Vector exchange, 37
Vehicle networks, 5
Virtual carrier sensing, 6, 24, 44

W

Weight vector adjustment, 5
Wireless links, 3
Wireless networks. *See also* Ad hoc networks
    factors affecting performance of, 41

Z

Zone, definition of, 138
Zone reservation, 137, 139, 185
    effectiveness of protocol for, 144
    high-priority flows, 176
Zone-disjoint routes, 101, 103, 109, 137
    multipath routing, 119
    path selection for, 140
    shortest path, 113, 116
    use with directional antennas, 121